Practice Papers for SQA Exams

Advanced Higher

Mathematics

Text © 2010 Edward Mullan
Design and layout © 2010 Leckie & Leckie

02/070211

ISBN 978-1-84372-860-3

Published by
Leckie & Leckie Ltd
An imprint of HarperCollins*Publishers*
Westerhill Road, Bishopbriggs, Glasgow, G64 2QT
T: 0844 576 8126   F: 0844 576 8131
leckieandleckie@harpercollins.co.uk   www.leckieandleckie.co.uk

A CIP Catalogue record for this book is available from the British Library.

Questions and answers in this book do not emanate from SQA. All of our entirely new and original Practice Papers have been written by experienced authors working directly for the publisher.

Leckie & Leckie makes every effort to ensure that all paper used in its books is made from wood pulp obtained from well-managed forests, controlled sources and recycled wood or fibre.

Mixed Sources
Product group from well-managed forests and other controlled sources
www.fsc.org   Cert no. SW-COC-001806
© 1996 Forest Stewardship Council

FSC is a non-profit international organisation established to promote the responsible management of the world's forests. Products carrying the FSC label are independently certified to assure consumers that they come from forests that are managed to meet the social, economic and ecological needs of present and future generations.

Find out more about HarperCollins and the environment at
www.harpercollins.co.uk/green

Introduction ... IMPORTANT!

Essentially the book contains three specimen papers, each assessing Advanced Higher Mathematics, Units 1, 2 and 3.

Each paper is balanced across the curriculum, sampling as much of the course as possible. A conscious effort to use a similar balance of topics and difficulty level as that seen in the SQA examinations has been made.

A Topic Index is included which will allow you to analyse any examination to look for this balance. This can also be used as a check-list when revising for the final exam.

In the case of differential equations, vectors, sequences and series, integration, the list of things to know can look quite 'bitty'. In an attempt to assist your revision, these four topics are summarized in an appendix.

In the case of vectors, it is useful to separate in your mind the **facts and skills** required from the **problem types**.

The book tries to expose you to as big a variety as possible. With that in mind you may feel there's a lot of work for the marks in certain cases. It is better to practice on material which is more demanding than that of the final exam. Aim high.

Athletes used to train for races by running along the strand. Running in shifting sand is a lot harder than running on the track and prepared them well for the real race.

Notes accompany the marking scheme. The working in many cases is more than you would be expected to produce in the final exam. This is because there is a **teaching/tutorial aspect** to the notes.

In some questions you are given information to help, information which may not be given in an actual exam. This may be formulae or techniques you learned in the Higher; it may be things you should know for the Advanced Higher.

In some cases the solutions hint at an extension to the question.

One cannot cover every approach to a topic that might be taken and this was felt a reasonable way to help you prepare for your exams. Don't just ignore a part of a solution just because it is not immediately relevant. Who knows what might be asked?

There is always more than one way of solving a problem. Sometimes a specific method is requested. Don't be tempted to use another. For example don't solve a system of equations by any other method if Gaussian elimination is asked for; don't integrate 'by parts' if a substitution method is specifically asked for; don't prove a statement true by direct methods if proof by induction is requested.

Certain topics seem to be represented more than others in the three papers.

This is because we wish to expose you to as wide a selection of questions as possible and some topics, e.g. Differential equations have a bigger variety of question types than others. The use of Maclaurin's series has only been examined once within the book.

While doing this we have made sure that the balance within a paper is maintained.

Note

For the sake of clarity, a dot is often used instead of the usual multiplication sign.

Such a dot will appear on the line so as not to be confused with a decimal point, which is mid-line.

e.g. 3·4 is read as '3 point 4' whereas 3.4 is read as '3 times 4'

The table below gives a broad idea of the coverage of the syllabus by the specimen papers.

Topic Index

UNIT 1	Exam 1	Exam 2	Exam 3	Knowledge for Prelim			Knowledge for SQA Exam		
				Have difficulty	Still needs work	OK	Have difficulty	Still needs work	OK
Factorials, Binomial Theorem, Partial Fractions									
Factorial function	5	14a							
Permutation and combination									
Binomial theorem	5	1a							
Approximation		1b							
rational functions and division	8b		8,11						
Distinct linear factors									
Repeated linear factors	6								
Improper rational functions									
Differential Calculus									
first principles	1								
Limits									
sum and difference rule									
chain rule	1a	2a	1a						
product rule	1a		1a						
quotient rule	1b								
sec,cosec, tan, cotan	1b								
exponential and log	1a,10								
higher derivatives									
rectilinear motion									
critical points and extrema			5,7						
Optimization									
other uses [e.g. newton-raphson]									
Integral Calculus									
Antiderivatives			3b						
standard form sin, cos, e^x, x^{-1}, $sec^2 x$	6	6	3a						
Differentials			3a						
integration by substitution	4	6	3a						
substitution and definite integrals			3,13						
particular cases to remember		6							
area between curves									
volumes of revolution		12							
rectilinear motion									
Properties of Functions									
modulus function									
curve sketching	9f		5						
Concavity	9e		7						
odd and even functions	9a								
Asymptotes	9bcd		5						
related functions									
System of Equations									
using matrices on a 2 × 2; augmented matrix	2	3	2						
EROs and 3 × 3 system	2	3	2						
Gaussian elimination	2	3	2						
redundancy and inconsistency			2						
ill-conditioning									

UNIT 2	Exam 1	Exam 2	Exam 3	Knowledge for Prelim			Knowledge for SQA Exam		
				Have difficulty	Still needs work	OK	Have difficulty	Still needs work	OK
Elementary Number Theory									
existential, universal, example, counterexample			6a						
direct proof		5b							
indirect proof			6a						
proof by induction	3	5a							
fundamental theorem of arithmetic									
Further Differential Calculus									
derivative of inverse functions		2a							
inverse trig functions		4							
implicit and explicit functions		2b	1,5						
logarithmic differentiation									
applications: motion in a plane		8							
applications related rates of change									
Parametric Equations	7		7						
Further Integration									
find intersection of two poly^{ls}									
standard integrals involving inverse trig fns	6	6							
rational functions	6	6							
integration by parts	4	12	3a						
differential equations (variables separable)	10	9	8						
applications of differential equations.	10		8						
Complex Numbers									
addition, subtraction, multiplication	8	10							
division and square root		10ab							
argand diagrams			4						
loci on the complex plane									
polar form and multiplication			4						
DeMoivre's theorem			4						
Roots of a complex number and polynomials	8		4						
Sequences and Series									
Recurrence relations									
Arithmetic Sequences	16	11a							
Geometric Sequences									
Infinite series, partial sums and sum to infinity		11bc	11						
Expanding $(1 - x)^{-1}$			11						
Sigma notation			11						

UNIT 3	Exam 1	Exam 2	Exam 3	Knowledge for Prelim			Knowledge for SQA Exam		
				Have difficulty	Still needs work	OK	Have difficulty	Still needs work	OK
Matrices									
Definitions and operations	13	7	9						
Algebraic properties	13	7	9						
matrix multiplication		7	9						
determinant [2 × 2 and 3 × 3]	13a								
inverse [2 × 2 and 3 × 3]	13b	7	9						
transformation matrices		7							
composition of transformations		7							
Vectors									
right and left handed systems									
vector product	15a	13b	10						
scalar triple product									
Equation of a plane	15a	13b	10b						
Problems: coplanar vectors	15c		10b						
Problems: angle between planes									
Problems: distance between parallel planes									
Equation of a line	15b	13a	10						
Problems: angle between a line and a plane		13c	10d						
Problems: intersection of two lines			10a						
Problems: intersection of two planes	15b	13							
Problems: intersection of a line and a plane									
Problems: distance from a point to a plane									
Problems: distance from a point to a line									
Problems: intersection of 3 planes									
Further Sequences and Series									
Power Series		14							
convergence			11						
Maclaurin series		14a							
Composite functions		14b							
Iterative processes									
Further Differential Equations									
Initial conditions, general and particular solns	12	16	8,12						
1st order linear variables separable	10	9							
1st order linear integrating factor	12		12						
2nd order linear diff. equations (homogeneous)	14	16	15						
2nd order non-homogeneous	14	16	15						
Further Number Theory									
Further direct and indirect proof		5b	14c						
Further proof by induction			6b						
Division algorithm	11a	15	14						
Euclidean algorithm	11a	15a	14						
Diophantine Equations	11b	15b	14						
Number bases									

Solving differential Equations

1 $\dfrac{dy}{dx} = f(x)$

Integrate to get y. $y = \displaystyle\int f(x)\, dx$

2 $\dfrac{dy}{dx} = f(x)g(y)$

Separate the variables and integrate. $\displaystyle\int \dfrac{dy}{g(y)} = \int f(x)\, dx$

3 $\dfrac{dy}{dx} + P(x)y = Q(x)$

Multiply by integrating factor, $I = e^{\int P(x)dx}$ and integrate to get

$Iy = \displaystyle\int IQ(x)\, dx$ and tidy to make y the subject.

4   Homogeneous Second order

$a\dfrac{d^2y}{dx^2} + b\dfrac{dy}{dx} + cy = 0$

Form the auxiliary equation $am^2 + bm + c = 0$ and solve it

(i)   If roots are real and distinct, the solution is $Ae^{m_1 x} + Be^{m_2 x}$

(ii)   If roots are real and coincident, the solution is $Ae^{mx} + Bxe^{mx}$

(iii)   If roots are complex conjugates, $m = p \pm qi$, the solution is $e^{px}(A\cos qx + B\sin qx)$

5   Non-Homogeneous Second order

$a\dfrac{d^2y}{dx^2} + b\dfrac{dy}{dx} + cy = Q(x)$

The solution = complementary function (CF) + particular integral (PI).

The CF is the solution for the corresponding homogeneous equation.

The particular integral is spotted by considering

(i)   general functions of the form $y = Q(x)$ e.g. $Q(x) = 3x$ we try $Cx + D$ (see p 125)

(ii)   if that ends up the same form as a term in the CF then try $y = xf(x)$.

(iii)   if that ends up the same form as a term in the CF then try $y = x^2 f(x)$.

We test PIs by working out y, $\dfrac{dy}{dx}$ and $\dfrac{d^2y}{dx^2}$ and substituting in the equation.

This will let us work out the values of C and D ... by equating coefficients.

name	formula	note
magnitude	$\|a\| = \sqrt{a_1^2 + a_2^2 + a_3^2}$	
Scalar product	$a.b = \|a\|.\|b\|.\cos\theta = a_1 b_1 + a_2 b_2 + a_3 b_3$	Good for finding θ
Vector product	$a \times b = n\|a\|.\|b\|.\sin\theta = \begin{vmatrix} i & j & k \\ a_1 & a_2 & a_3 \\ b_1 & b_2 & b_3 \end{vmatrix}$	Gets you a vector at right angles to both a and b … *use right hand rule.*
Scalar triple product	$a.(b \times c) = \begin{vmatrix} a_1 & a_2 & a_3 \\ b_1 & b_2 & b_3 \\ c_1 & c_2 & c_3 \end{vmatrix}$	Gets you the volume of the parallelopiped with edges a, b, c.
Equation of a plane	For any plane, the scalar product of a (a normal to the plane) and x (the position vector of any point on the plane) is a constant: $a \cdot x = k$ $$\begin{pmatrix} a \\ b \\ c \end{pmatrix} . \begin{pmatrix} x \\ y \\ z \end{pmatrix} = k \Rightarrow ax + by + cz = k$$ e.g. $3x + 2y - 4z = 1$ is a plane with a normal $\begin{pmatrix} 3 \\ 2 \\ -4 \end{pmatrix}$ or $3i + 2j - 4k$. Check that $(5, -1, 3)$ lies on the plane and $(1, 2, 1)$ does not.	If we know a point and a normal we can find k and then quote the equation for the general point (x, y, z). If we know 2 points we can state a vector on the plane; with 3 points we have 2 vectors; their vector product provides a normal … so we now have a normal and a point …
Equation of a line	If r is the position vector of any point, R, on a line and a is the position vector of a fixed point, A, on the line and u is a vector parallel to the line then $\overrightarrow{AR} = r - a = tu$ where t is a constant. This is the equation of the line. It is written in 3 forms : $r = a + tu$ … the vector form $$\begin{pmatrix} x \\ y \\ z \end{pmatrix} = \begin{pmatrix} x_1 \\ y_1 \\ z_1 \end{pmatrix} + t \begin{pmatrix} a \\ b \\ c \end{pmatrix}$$ which becomes the system: $x = x_1 + at$ $y = y_1 + bt$ $z = z_1 + ct$ … parametric form solving each for t we get the symmetric form $$\frac{x - x_1}{a} = \frac{y - y_1}{b} = \frac{z - z_1}{c} \quad (= t)$$	You should be able to switch between forms. To change from the symmetric form you have to reintroduce the '$= t$'. Given a line, you should be able to pick out a point on the line (x_1, y_1, z_1) and a parallel vector $ai + bj + ck$ You need a direction vector and a point on the line; using two points on the line you can get a direction vector.

Vectors Common Strategies	
Angle between two planes	This is the same as the angle between their normals: • read the two normals from their equations • use the scalar product to find the angle between them.
Angle between two lines	This is the same as the angle between their direction vectors: • read direction vectors from equations • use the scalar product to find the angle between them
Angle between a line and a plane	This is the complement of the angle between the direction vector of the line and the normal to the plane.
Intersection of a line and a plane.	• Express the line in parametric form. • Substitute these expressions for x, y and z into the equation of the plane. • Solve for t. • Substitute this value of t into the expressions for x, y and z.
Intersection of two lines.	• Express both lines in parametric form using t_1 and t_2 • equate expressions for x, y and z creating a system of 3 equations with 2 variables. • use two of these equations to find t_1 and t_2 • substitute these values into 3rd equation. If they satisfy it then we have the point of intersection. If they don't then there is no intersection.
Intersection of two planes.	• If planes intersect they form a line. • find a point on the line by setting $z = 0$ on both planes and solve the resulting system … this should give a point on the desired line. (if it doesn't, go back and set $x = 0$ instead) • find the vector product of the normals to the planes … this will be a direction vector to the line. • you now have a point and a direction and hence the equation.

Intersection of three planes.	• Solve the 3×3 system of equations
	  (i) unique answer … point of intersection
	  (ii) one equation redundant … line of intersection.
	  (iii) inconsistent … 2 or 3 lines … examine the planes in pairs
	  (iv) two redundant … planes coincident.
	  (v) two inconsistent lines … planes parallel

Distance between a point, P, and a plane	Consider the point P′, which lies on the plane such that PP′ is perpendicular to plane. • Find equation of PP′ … parallel to normal of plane and passing through P. • Find P′, where the found line intersects the given plane. • Find the distance PP′ using distance formula. P′ is known as the projection of P on the plane. You may be asked for this. [bullets 1 and 2]

Vectors Common Strategies	
Distance between a point, P, and a line	Consider the point P′, which lies on the line such that PP′ is perpendicular to line. • Express P′ in terms of t the parameter of the line • Hence express PP′ in terms of t. • The scalar product of PP′ and direction vector of line = 0 (since PP′ perpendicular to line). Use this to find t corresponding to P′. • Hence find P′ and PP′. P′ is known as the projection of P on the line.
Distance between parallel planes	The distance between $ax + by + cz = m$ and $ax + by + cz = n$ is given by the formula: $$d = \frac{|m - n|}{\sqrt{a^2 + b^2 + c^2}}$$ note that $\begin{pmatrix} a \\ b \\ c \end{pmatrix}$ is the common normal.

Sequences and Series

1 1st order linear recurrence relations

These are of the form $u_{n+1} = ru_n + d$ where r and d are constants.

2 An arithmetic sequence ... is generated when $r = 1$.

By convention $u_1 = a$ and d is called the common difference.

(a) $u_n = a + (n-1)d$ ... proof by induction

(b) $S_n = \frac{n}{2}(u_1 + u_n)$ ... direct proof

$S_n = \frac{n}{2}(2a + (n-1)d)$

3 A geometric sequence ... is generated when $d = 0$.

By convention $u_1 = a$ and r is called the common ratio.

(a) $u_n = ar^{n-1}$ ... proof by induction

(b) $S_n = \frac{a(1-r^n)}{(1-r)}$ ... direct proof

(c) $S_\infty = \frac{a}{1-r}, -1 < r < 1$

4 Sigma Notation

(a) $\displaystyle\sum_{r=1}^{n} f(r) = f(1) + f(2) + f(3) + \cdots + f(n)$

(b) $\displaystyle\sum_{r=1}^{n} 1 = 1 + 1 + 1 + \cdots + 1 = n$

(c) $\displaystyle\sum_{r=1}^{n} r = 1 + 2 + 3 + \cdots + n = \frac{n}{2}(n+1)$

(d) $\displaystyle\sum_{r=1}^{n}(ar + b) = a\sum_{r=1}^{n} r + b\sum_{r=1}^{n} 1$

(e) $\displaystyle\sum_{r=1}^{n} r^2 = 1^2 + 2^2 + 3^2 + \cdots + n^2 = \frac{n}{6}(n+1)(2n+1)$

(f) $\displaystyle\sum_{r=1}^{n} r^3 = 1^3 + 2^3 + 3^3 + \cdots + n^3 = \frac{n^2}{4}(n+1)^2 = \left(\sum_{r=1}^{n} r\right)^2$

5 Maclaurin's series

$$f(x) = \sum_{r=0}^{\infty} \frac{f^{(r)}(0)}{r!}x^r \text{ where } f^{(r)} \text{ is the } r^{th} \text{ derivative and } f^{(0)}(0) = f(0)$$

6 Taylor's series

$$f(x + h) = \sum_{r=0}^{\infty} \frac{f^{(r)}(x)}{r!}h^r$$

7 Iterative processes (Newton-Raphson as an example)

Solving $f(x) = 0$, if x_n is guess then x_{n+1} is a better guess and $x_{n+1} = x_n - \dfrac{f(x_n)}{f'(x_n)}$

Standard Integrals for Advanced Higher

ref	function	integral
1	x^n	$\dfrac{x^{n+1}}{n+1},\quad n \neq 1$
2	$\dfrac{1}{x}$	$\log_e|x|$
3	e^x	e^x
4	$\log_e x$	$x\log_e x - x$
5	$\sin x$	$-\cos x$
6	$\cos x$	$\sin x$
7	$\tan x$	$-\log_e(\cos x)$

ref	function	integral
8	$\sec^2 x$	$\tan x$
9	$\dfrac{1}{\sqrt{a^2 - x^2}}$	$\sin^{-1}\left(\dfrac{x}{a}\right)$ or $-\cos^{-1}\left(\dfrac{x}{a}\right)$
10	$\dfrac{1}{a^2 + x^2}$	$\dfrac{1}{a}\tan^{-1}\left(\dfrac{x}{a}\right)$
11	$\dfrac{1}{\sqrt{x^2 \pm a^2}}$	$\log_e\left|x + \sqrt{x^2 \pm a^2}\right|$

The above list can be extended using the fact that if

$$\int f(x)\,dx = F(x) \text{ then } \int f(ax + b)\,dx = \frac{1}{a}F(ax + b)$$

Look out for cases of $\dfrac{f'(x)}{f(x)}$, for $\displaystyle\int \frac{f'(x)}{f(x)}\,dx = \log_e\left|f(x)\right|$

Substitution

If you have an integral of the form $\displaystyle\int ag'(x)f(g(x))\,dx$ where a is a constant,

you can let $g(x) = u$ and so $g'(x)dx = du$ and the integral simplifies to $\displaystyle\int af(u)\,du$

By Parts

If you are asked to integrate the product of functions, say uv where u and v are functions of x:

$$\int uv\,dx = u\int v\,dx - \int\int v\,dx \cdot \frac{du}{dx}\,dx$$

and you hope the new integral is easier than the one with which you started.

- Often the function which becomes simpler when differentiated is chosen for u.

- You may have to repeat the breakdown into parts more than once.

- If $f(x)$ cannot be directly integrated sometimes integrating $f(x) \cdot 1$ by parts, using $u = f(x)$ will work.

Exam 1

Advanced Higher Mathematics

Practice Papers for SQA Exams	Time allowed: 3 hours	Exam 1

You have 3 hours to complete the exam.

You can use a calculator in this exam.

Try to answer all of the questions in the time allowed.

Remember to include all appropriate working in your answer.

Scotland's leading educational publishers

Marks

1. Differentiate with respect to x

   (a) $(1-x^2)\ln(\cos x)$ — 4

   (b) $\dfrac{\tan x}{\sqrt{x}}$ — 2

2. Use Gaussian elimination to solve the following system of equations.

$$\begin{aligned}
x &-2y &+3z &= 7 \\
2x &+3y &+z &= 2 \\
3x &+y &-z &= 4
\end{aligned}$$

   5

3. Prove by induction that for all natural numbers n,

$$\frac{1}{1.2} + \frac{1}{2.3} + \frac{1}{3.4} + \cdots + \frac{1}{n.(n+1)} = \frac{n}{n+1}$$

   5

4. Evaluate

$$\int_0^{\pi/4} 3x\cos 2x \, dx$$

   5

5. Use the binomial theorem to find the coefficient of the term independent of x in the expansion of $\left(x^2 + \dfrac{3}{x^4}\right)^9$

   5

6. Use partial fractions to help you integrate the function $f(x) = \dfrac{2x^2 + 2x + 5}{(2x+3)(x^2+1)}, \; x > 0$

   6

7. The curve shown has parametric equations

   $x = 4 + 2\sin(2\theta)$

   $y = 3 + 2\cos(\theta)$

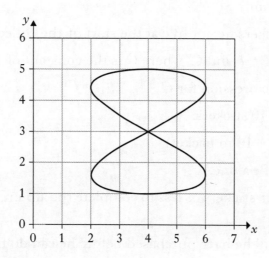

   (a) Find $\dfrac{dy}{dx}$ in terms of θ.

   2

   (b) The curve crosses itself when $y = 3$.

   Calculate the values of θ for which this occurs, $0 \le \theta < 2\pi$

   1

   (c) Obtain the equations of the tangents for these values of θ.

   2

8. The equation $z^4 - 8z^3 + 39z^2 - 62z + 50 = 0$ has a root $z = 3 + 4i$.

    (*a*) State one other root.   **1**

    (*b*) Find all the roots of the equation.   **4**

9. A function f is defined over a suitable domain by $f(x) = \dfrac{x^3}{x^2 - 4}$

    (*a*) Prove that it is an odd function.   **2**

    (*b*) Find the equations of the vertical asymptotes.   **1**

    (*c*) Find the equation of the non-vertical asymptote.   **2**

    (*d*) Calculate the point at which the curve cuts this non-vertical asymptote.   **2**

    (*e*) Calculate the *x*- coordinates of the stationary points leaving them in surd form and determine their nature.   **3**

    (*f*) Sketch the curve.   **1**

10. Mike is collecting football stickers to put in an album.

There are 500 to collect.

He buys the album and gets some free stickers with it.

These initial stickers cannot be obtained any other way.

He buys a packet of 10 stickers each day, discards any stickers he may already have and puts the rest in the album.

The size of his collection can be modelled by the differential equation

$$\frac{dN}{dt} = k(500 - N)$$

where $N(t)$ is the number of stickers he has at time t days after he started his collection and k is a constant.

$N(0)$ is the number of stickers he got free at the start of the collection.

    (*a*) Express N in terms of t, k and C, where C is the constant of integration.   **3**

    (*b*) If $N(0) = 10$ find an expression for C.   **1**

    (*c*) Each packet contains 10 stickers.

        ($N(1) = 20$ ... 10 free + 10 in packet)

        Express N in terms of t alone.   **1**

    (*d*) When he needs only 25 stickers or less to complete the album, he can send away for them.

        How many packets will he have purchased before he can do this?   **2**

11. (*a*) Use the Euclidean Algorithm to obtain the greatest common divisor of 3255 and 4785.   **2**

    (*b*) Hence find two integers *x* and *y* such that $3255x + 4785y = 360$ **3**

Marks

12. Find the particular solution to the differential equation

$$\frac{dy}{dx} + 3x^2y = (1 + 3x^2)e^x \text{ given that when } x = 0, y = 1.$$

5

13. The matrix is defined by $A = \begin{pmatrix} 4 & 1 & x \\ -5 & -1 & 2 \\ -2 & 0 & y \end{pmatrix}$

    (a) Find the relation between x and y that would make A singular. **2**

    (b) For particular values of x and y the inverse of A is $A^{-1} = \begin{pmatrix} -3 & -3 & 1 \\ 11 & 10 & -3 \\ -2 & -2 & 1 \end{pmatrix}$

    Find these values. **3**

14. Find the general solution of the differential equation

$$\frac{d^2y}{dx^2} + 7\frac{dy}{dx} + 12y = 4x(6x + 1)$$

5

15. A plane, π_1, passes through $(-1, 2, 4)$, $(1, 1, -4)$ and $(-2, 3, 7)$.

    (a) Find the equation of the plane **4**

    (b) A second plane π_2 has equation $2x + y + 3z = 1$.

    Find the equation of the line which is formed by the intersection of the two planes. **3**

    (c) Calculate the angle between the planes. **3**

16. Two different arithmetic sequences when summed to 21 terms come to the same amount, 1176.

    (a) The first sequence starts with 6.

    What is the second term of the sequence? **2**

    (b) The second sequence starts with a and has a common difference of d.

    Express a in terms of d. **2**

    (c) The ninth term of the second sequence is 50, Find the first and second terms. **3**

    (d) For some particular value of n, the nth terms of the two sequences are the same.

    Find this term. **3**

[End of question paper]

Exam 2

Advanced Higher Mathematics

Practice Papers for SQA Exams	Time allowed: 3 hours	Exam 2

You have 3 hours to complete the exam.

You can use a calculator in this exam.

Try to answer all of the questions in the time allowed.

Remember to include all appropriate working in your answer.

Leckie ✕ Leckie

Scotland's leading educational publishers

Marks

1. (a) Expand $\left(x + \frac{1}{x}\right)^4$ using the binomial theorem and simplify.   2

   (b) Hence find the value of $10\cdot1^4$ to the nearest whole number.   3

2. Find the derivative of

   (a) $\sec^2 x$ 2

   (b) 3^{2x} 3

3. A circle has an equation of the form $x^2 + y^2 + 2gx + 2fy + c = 0$.

   It passes through the points $(-1, 1)$, $(7, -3)$ and $(0, 4)$.

   (a) Form a system of equations in g, f, and c.   1

   (b) Use Gaussian elimination to find the values of g, f and c and thus the equation of the circle.   5

4. $x^2 + y^2 + xy = 9$ is the equation of an ellipse as shown.

Ellipse

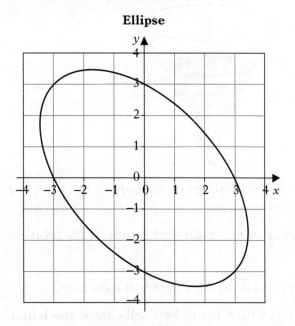

   At what points on the ellipse is the gradient

   (a) 1   3

   (b) 0   2

   (c) undefined?   1

5. (a) Prove by induction that $x^2 + 2x - 4 > 0$ for all integral values of x bigger than 1.   5

   (b) Use a direct proof to arrive at the same conclusion.   2

Marks

6. A function is defined by $f(x) = \dfrac{2x + 1}{x^2 - 2x + 10}$; $x \in \mathbb{R}$

   (a) Express the denominator in the form $(x + a)^2 + b$ where a and b are constants. **1**

   (b) By using the substitution $u = x + a$ find $\displaystyle\int \dfrac{2x + 1}{x^2 - 2x + 10}\,dx$ **5**

7. B is a transformation matrix where $B = \begin{pmatrix} \dfrac{1}{\sqrt{2}} & -\dfrac{1}{\sqrt{2}} \\ \dfrac{1}{\sqrt{2}} & \dfrac{1}{\sqrt{2}} \end{pmatrix}$

   (a) Find B^2 and interpret it geometrically. **2**

   (b) Find B^4 and hence an expression in B for B^{-1}. **2**

   (c) A is a second transformation matrix where $A = \begin{pmatrix} \sqrt{2} & \sqrt{2} \\ -\sqrt{2} & \sqrt{2} \end{pmatrix}$

   Find a geometric interpretation for the matrix AB. **2**

8. A conical hopper is 2 metres tall with a radius of 50 cm and filled with water.

   It is emptying at a constant rate of 100 ml per second.

   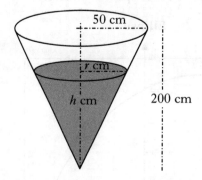

   (a) Express the volume V, ml, of water in the hopper in terms of r when r cm is the radius of the surface of the water. **3**

   (b) Find the rate of change of this radius with time at the moment when it is 20 cm. **4**

9. Tickets for a concert go on sale 30 days before the concert.

   Over the month the rate at which the tickets sell can be modelled by

   $$\frac{dP}{dt} = \frac{t^2}{100}(100 - P)$$

   where the percentage of tickets sold at time t days after the sale begins is P%.

   Obviously when $t = 0$, $P = 0$.

   Express P in terms of t explicitly **6**

10. (a) Find x and y such that $(4 + 3i)(x + iy) = 1 - i$ **3**

    (b) Find a and b such that $a + ib = \sqrt{(5 + 12i)}$ **3**

Marks

11. The first term of a geometric series is 810. The fifth term is 10

    (a) How do you know a sum to infinity exists?   **1**

    (b) The sum to infinity is 1215. List the first five terms.   **3**

    (c) Which term is the first one less than 0·01?   **2**

12. When the graph of the function $y = x^2\sqrt{(\ln x)}$ is rotated about the y-axis, a 'saucer-shape' is formed.

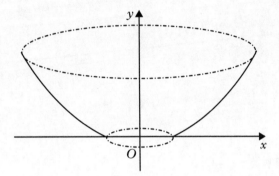

The volume of the shape can be calculated by evaluating $V = \pi\int_1^4 y^2\, dx$

Calculate the volume of the saucer.   **6**

13. A line lies on a plane, π, and passes through the points $(1, -1, 3)$ and $(3, 1, 4)$.

    (a) Find the equation of the line in symmetric form.   **4**

    (b) The point $(-1, 2, 0)$ also lies on this plane.

         Find the equation of the plane.   **3**

    (c) What angle does the line with parametric equations

         $x = 3t + 1$, $y = t - 4$, $z = 5t + 3$ make with the plane π?   **3**

14. (a) Write down the first four non-zero terms of Maclaurin's expansion for $\sin x$.   **1**

    (b) Hence state a power series for $\sin 2x$, again giving the first four non-zero terms.   **1**

    (c) Find a power series for $\sin\left(2x + \dfrac{\pi}{3}\right)$.   **4**

[You may find it useful to know $\sin(A \pm B) = \sin A\cos B \pm \cos A\sin B$]

15. (a) Find the greatest common divisor of 210 and 198 using the Euclidean algorithm.   **2**

    (b) Hence find a point on the line with equation $198x + 210y = 3090$ with integer co-ordinates.   **2**

    (c) Prove that if (a, b) is such a point then so is $(a + 210k, b - 198k)$.   **2**

Marks

16. Find the particular solution to the differential equation

$$\frac{d^2y}{dx^2} - 8\frac{dy}{dx} + 16y - 16x + 4 = 0 \text{ given that}$$

when $x = 0$, $y = \frac{5}{4}$ and $\frac{dy}{dx} = 0$.

6

[End of question paper]

Advanced Higher Mathematics

Practice Papers for SQA Exams	Time allowed: 3 hours	Exam 3

You have 3 hours to complete the exam.

You can use a calculator in this exam.

Try to answer all of the questions in the time allowed.

Remember to include all appropriate working in your answer.

Leckie✕Leckie
Scotland's leading educational publishers

Marks

1. Find the derivative of

    (a) $(2x + 1)\cos^2 x$ 3

    (b) x^x 3

2. A system of equations is given where x, y and z are variables and t is a parameter:

    $x + y + z = 2$

    $2x - y + (2t - 13)z = -9$

    $4x + 3y - z = -3$

    Use Gaussian elimination to determine the solution to the system in terms of t and state any necessary conditions.   4

    Show that there is no value of t for which all three numbers, x, y and z are positive.   1

3. (a) By considering the substitution $x = 3\cos u$, evaluate

    $$\int_0^{\frac{3}{2}} \frac{x\cos^{-1}\frac{x}{3}}{\sqrt{9 - x^2}}\,dx$$ 5

    (b) (i) Expand and simplify $\frac{1}{2}\big(\sin(A + B) + \sin(A - B)\big)$

    (ii) Hence integrate $\sin 3x \cos 5x$ with respect to x.   3

4. Find all the solutions to the equation $z^5 = 32$ expressing the solutions in the form $z = a + ib$ where a and b are real numbers.   3

5. A curve is defined by $y = \dfrac{x^5}{(x + 1)^3(x - 1)^2}; x \neq \pm 1$

    (a) Use logarithmic differentiation to find $\dfrac{dy}{dx}$ and hence the stationary points and their nature of the curve.   4

    (b) State the equations of the vertical asymptotes and, by considering the behaviour of y as $x \to \infty$, the non-vertical asymptote.   2

    (c) By considering the value of y at $x = 0 \cdot 7$ and $x = 0 \cdot 8$ say why the curve cuts its own non-vertical asymptote.   1

    (d) Given that it cuts its own non-vertical asymptote again between $x = 2 \cdot 5$ and $x = 2 \cdot 6$, make a sketch of the curve.   1

Marks

6. (a) It has been conjectured that $n^2 + n + 41$ is a prime number for all n, $n \in N$. Prove the conjecture is wrong.

1

(b) Prove by induction that for each $n > 5$, $n \in N$, there exists whole numbers a and b such that $2a + 7b = n$

4

7. For Valentine's day a mathematician sent the following message:

"Love and Kisses, $x = 2\sin\theta + \sin 2\theta$; $y = -2\cos\theta - 2\cos^2\theta$; $0 \le \theta < 2\pi$."

The parametric equations define a pertinent curve which is symmetrical about the y-axis.

(a) Establish where the curve cuts the axes.

2

(b) Find $\dfrac{dy}{dx}$ and hence:

    (i) find where the gradient is zero;

3

    (ii) find where the gradient is undefined;

2

(c) Given that the curve is concave down for $y > -1 \cdot 5$ and concave up for $y < -1 \cdot 5$, sketch the curve.

1

8. Making preparations to guard against environmental problems in Loch Lomond, scientists model the spread of a potential algal bloom by the equation:

$$\frac{dH}{dt} = \frac{H}{100\,000}(7000 - H)$$ where H hectares is the area covered by the bloom t hours after it is discovered.

The model assumes that when the bloom is first detected ($t = 0$) there will be 100 hectares affected.

(a) Express H explicitly in terms of t.

6

(b) Calculate what area the model estimates that the bloom will cover after 24 hours.

1

(c) Given that Loch Lomond has an area of 7000 hectares, how long will it be before the bloom affects 90% of the Loch?

2

9. Given the 3×3 matrix $A = \begin{pmatrix} 4 & 1 & 1 \\ 1 & 0 & 2 \\ 3 & 1 & 0 \end{pmatrix}$,

(a) Use elementary row operations to find A^{-1}.

3

(b) Hence find the matrix B where $AB = \begin{pmatrix} 4 & 3 & 8 \\ 9 & 5 & 1 \\ 2 & 7 & 6 \end{pmatrix}$ the entries of which form a magic square.

2

(c) Show that this matrix, B, is non-singular.

2

Marks

10. The equations $\dfrac{x+1}{2} = \dfrac{y+1}{3} = \dfrac{z}{3}$ and $\dfrac{x+8}{3} = \dfrac{y+10}{4} = \dfrac{z+3}{2}$ define two intersecting lines in space.

   (a) Find their point of intersection.   **4**

   (b) Find the equation of the plane upon which these two lines lie.   **3**

   (c) A normal to the plane passes through the origin.

   Find the equation of this line, expressing it in parametric form.

   Calculate the value of the parameter at the point where the line cuts the plane.   **2**

   (d) Calculate the cosine of the angle this normal makes with each of the axes.   **1**

11. R_n is defined by the formula: $\dfrac{1}{R_n} = \dfrac{1}{2} + \dfrac{1}{6} + \dfrac{1}{12} + \cdots + \dfrac{1}{n^2 + n}$

   (a) Express the right hand side of the equation using sigma notation.   **1**

   (b) By factorizing the denominator and using partial fractions express R_n as a single fraction in terms of n.   **3**

   (c) In a similar manner, express $\displaystyle\sum_{r=2}^{n} \dfrac{1}{(r^2 - 1)}$ in terms of n alone.

   Hence, find the limit of the sum as n tends to infinity.   **3**

   [Note: Be careful. The lower limit to r is 2 and not 1]

12. Solve $\cos x \dfrac{dy}{dx} + y \sin x = x$, expressing y explicitly in terms of x, and find the particular solution if, when $x = 0$, $y = 1$.   **6**

13. The length of a curve, L units, from $x = a$ to $x = b$ can be calculated using the formula $L = \displaystyle\int_{a}^{b} \sqrt{1 + \left(\dfrac{dy}{dx}\right)^2}\, dx.$

   The following diamond-shaped curve has parametric equations

   $$x = \cos^3\theta; \; y = \sin^3\theta; \; 0 \le \theta < 2\pi$$

   Diamonds

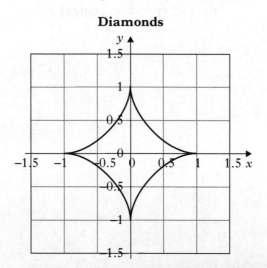

13. (continued)

(a) Express dx in terms of θ.   **1**

(b) Express $\dfrac{dy}{dx}$ in terms of θ.   **1**

(c) By considering $L = \displaystyle\int_a^b \sqrt{1 + \left(\dfrac{dy}{dx}\right)^2}\, dx$ for $0 \le \theta \le \dfrac{\pi}{2}$, find the perimeter of the shape.   **4**

14. The least common multiple, $L(a,b)$ of two numbers a and b, can be calculated from the formula $L(a,b) = \dfrac{ab}{G(a,b)}$ where $G(a,b)$ represents the greatest common divisor of a and b.

The largest asteroid Ceres has an orbital period, C hours, where $C = 40\ 152$. The second largest, Vesta has a period, V hours, where $V = 31\ 752$.

On a particular date the distance between the two asteroids is at its smallest.

(a) An astronomer sets up a camera programmed to capture that part of the heavens at regular intervals.

He decides that this interval should be G(40 152, 31 752)

What is the interval between photos?

Make use of the Euclidean algorithm.   **3**

(b) After how many hours will the distance be again at its smallest?   **1**

[Hint: Find L(40 152, 31 752)]

(c) Find integers a and b such that $aC + bV = 168$ and show that $\dfrac{1}{L(C,V)} = \dfrac{a}{V} + \dfrac{b}{C}$ **3**

15. Find the general solution to the equation

$$\frac{d^2 y}{dx^2} - 6\frac{dy}{dx} + 13y - 21\sin 2x = 0\,.$$
    5

[End of question paper]

Worked Answers

Answers to Practice Exam 1

SOLUTIONS, NOTES AND MARKING SCHEME FOR EXAM 1

1. *(a)* $y = (1 - x^2)\ln(\cos x)$

$$\Rightarrow \frac{dy}{dx} = (1 - x^2) \cdot \frac{1}{\cos x} \cdot -\sin x + \ln(\cos x) \cdot -2x$$

$$= -(1 - x^2)\tan x - 2x\ln(\cos x)$$

Marks   **1** for method (using product rule);

1 for method (using chain rule);

1 for processing first term;

1 for processing second term.

> **NOTES**   The question is testing your knowledge of both the product rule and the chain rule. Try to show by your intermediate working that you do indeed know these algorithms.

1. *(b)* $y = \dfrac{\tan x}{x^{\frac{1}{2}}}$

$$\Rightarrow \frac{dy}{dx} = \frac{x^{\frac{1}{2}}\sec^2 x - \tan x \cdot \frac{1}{2}x^{-\frac{1}{2}}}{x}$$

$$= \frac{2x\sec^2 x - \tan x}{2x^{\frac{3}{2}}}$$

Marks   **1** for method (using quotient rule);

1 for processing the derivative of $\tan x$.

> **NOTES**   The question is testing your knowledge of both the quotient rule and the derivative of $\tan x$. As long as you show these then the final 'tidying up' is not essential.

2.
$$\begin{pmatrix} 1 & -2 & 3 & 7 \\ 2 & 3 & 1 & 2 \\ 3 & 1 & -1 & 4 \end{pmatrix}$$

(i) $R2 \rightarrow R2 - 2R1$
(ii) $R3 \rightarrow R3 - 3R1$
$$\begin{pmatrix} 1 & -2 & 3 & 7 \\ 0 & 7 & -5 & -12 \\ 0 & 7 & -10 & -17 \end{pmatrix}$$

(iii) $R3 \rightarrow R3 - R2$
$$\begin{pmatrix} 1 & -2 & 3 & 7 \\ 0 & 7 & -5 & -12 \\ 0 & 0 & -5 & -5 \end{pmatrix}$$

Row three: $\Rightarrow -5z = -5 \Rightarrow \mathbf{z = 1}$

Row two: $\Rightarrow 7y - 5z = -12 \Rightarrow 7y - 5 = -12 \Rightarrow 7y = -7 \Rightarrow \mathbf{y = -1}$

Row one: $\Rightarrow x - 2y + 3z = 7 \Rightarrow x + 2 + 3 = 7 \Rightarrow \mathbf{x = 2}$

2. (continued)

Marks

1 for method (for structured method);

2 marks for manipulation to upper triangular form (lose one mark per error);

1 for value of z;

1 for values of x and y.

NOTES

Only a solution obtained by Gaussian Elimination is acceptable for marks.

The augmented matrix plus clear elementary row operations (EROs) to achieve upper triangular form are essential.

Thereafter back substitution is fine.

Indicate what operations you have chosen to perform. This will make checking your working afterwards easier.

The notation used here can best be described by an example:

$R2 \rightarrow R2 + 4R3$ is read as "Row 2 *becomes* Row 2 plus 4 times Row 3".

It should be noted that R2 refers to the most recent version of row 2.

The roman numeral lets you know the order in which the EROs are performed.

Remember that with each legitimate ERO you are creating a new system of equations with the same solution as the original system. (and you expect the new system to be simpler)

Some people prefer to work in matrices all the way to a solution:

So working upwards ...

$$(v)\ R2 \rightarrow R2 - R3 \quad (iv)\ R3 \rightarrow R3/7 \quad \begin{pmatrix} 1 & -2 & 3 & 7 \\ 0 & 7 & 0 & -7 \\ 0 & 0 & 1 & 1 \end{pmatrix}$$

$$(vii)\ R1 \rightarrow R1 - 3R3 \quad (vi)\ R2 \rightarrow R2/7 \quad \begin{pmatrix} 1 & -2 & 0 & 4 \\ 0 & 1 & 0 & -1 \\ 0 & 0 & 1 & 1 \end{pmatrix}$$

$$(viii)\ R1 \rightarrow R1 - 2R2 \quad \begin{pmatrix} 1 & 0 & 0 & 2 \\ 0 & 1 & 0 & -1 \\ 0 & 0 & 1 & 1 \end{pmatrix}$$

which gives $x = 2$, $y = -1$, $z = 1$

---

3. R.T.P. $\dfrac{1}{1.2} + \dfrac{1}{2.3} + \dfrac{1}{3.4} + \cdots + \dfrac{1}{n.(n+1)} = \dfrac{n}{n+1}$ $\forall\, n \geq 1,\, n \in N$

Consider the lowest case i.e. $n = 1$.

L.H.S. $= \dfrac{1}{1.2} = \dfrac{1}{2}$; R.H.S. $= \dfrac{1}{1+1} = \dfrac{1}{2}$.

L.H.S. = R.H.S. so the proposal is true for $n = 1$.

Suppose that it is true for some number $n = k$.

Then $\dfrac{1}{1.2} + \dfrac{1}{2.3} + \dfrac{1}{3.4} + \cdots + \dfrac{1}{k.(k+1)} = \dfrac{k}{k+1}$ $\cdots$ ①

3. (continued)

Consider when $n = k + 1$

$$\frac{1}{1.2} + \frac{1}{2.3} + \frac{1}{3.4} + \cdots + \frac{1}{k.(k+1)} + \frac{1}{(k+1).(k+1+1)}$$

$$= \frac{k}{k+1} + \frac{1}{(k+1).(k+2)} \text{ (using the supposition } ① \text{ above)}$$

$$= \frac{k(k+2)+1}{(k+1).(k+2)} = \frac{k^2+2k+1}{(k+1).(k+2)} = \frac{(k+1)^2}{(k+1).(k+2)}$$

$$= \frac{k+1}{k+2} = \frac{(k+1)}{(k+1)+1}$$

So the proposal is true for $n = k + 1$

If the proposal is true for $n = k$ then the proposal is true for $n = k + 1$.

The proposal is true for $n = 1$ so, by induction the proposal is true $\forall\, n \geq 1$, $n \in N$.

Marks **1** for showing proposal true for $n = 1$ (LHS = RHS);

1 for assuming true for $n = k$ and starting to consider $n = k + 1$;

1 for using the assumption correctly;

1 for establishing that true for $n = k$ implies true for $n = k + 1$;

1 for pulling it all together in a final conclusion.

NOTES

The presentation of a proof by induction is rigorous.

- Show that the proposal is true for the lowest case ... in this example $n = 1$, though this is not always what's required.

- Show that if it is true for $n = k$ then it follows that it is true for $n = k + 1$

- Draw a conclusion that with the above two bullet points demonstrated as true then, by induction the proposal is true for all n bigger than or equal to the lowest case.

4.

$$\int_0^{\frac{\pi}{4}} 3x\cos 2x\ dx = \left[3x.\frac{1}{2}\sin 2x - \int \frac{1}{2}\sin 2x.3\ dx \right]_0^{\frac{\pi}{4}}$$

$$= \left[\frac{3}{2}x\sin 2x + \frac{3}{4}\cos 2x \right]_0^{\frac{\pi}{4}}$$

$$\left[\frac{3}{2}.\frac{\pi}{4}\sin\frac{\pi}{2} + \frac{3}{4}\cos\frac{\pi}{2} \right] - \left[\frac{3}{2}.0\sin 0 + \frac{3}{4}\cos 0 \right]$$

$$= \frac{3\pi}{8}.1 + \frac{3}{4}.0 - 0 - \frac{3}{4} = \frac{3\pi}{8} - \frac{3}{4}$$

Marks **1** for method (establishing the 'parts');

1 for performing initial integration and showing it in both terms;

1 for performing second integration;

4. (continued)

1 for using limits correctly;

1 for evaluation.

> *NOTES*
>
> Integration by parts has been used.
>
> $$\int uv \, dx = u \int v \, dx - \int \left(\int v \, dx \right) \cdot \frac{du}{dx} dx$$
>
> Here the parts are $u(x) = 3x$ and $v(x) = \cos 2x$.
>
> The choice is made because $3x$ becomes simpler on differentiation and $\cos 2x$ can be
>
> integrated. The hope is that $\int \left(\int v \, dx \right) \cdot \frac{du}{dx} dx$ will be easier to integrate than $\int uv \, dx$.

5. The general term of the expansion of $(a + b)^9$ is $\binom{9}{n} a^n b^{9-n}$.

In this example $a = x^2$ and $b = \dfrac{3}{x^4}$

So you have $\binom{9}{n}(x^2)^n \left(\dfrac{3}{x^4} \right)^{9-n}$

$$= \binom{9}{n} x^{2n} \cdot 3^{9-n} \cdot x^{-36+4n}$$

$$= \binom{9}{n} x^{6n-36} \cdot 3^{9-n}$$

For the conditions asked for, you want the power of x to be 0.

i.e. you want $6n - 36 = 0$

This means you want $n = 6$.

When $n = 6$, the term is $\binom{9}{6} x^0 . 3^{9-6} = \dfrac{9!}{6!.3!} . 1.3^3 = 2268$

Marks   **1** for stating the general term;

1 for powers;

1 for condition that term is independent of x;

1 for finding $n = 6$;

1 for consistent evaluation.

> *NOTES*   Remember the whole expansion is the sum of all the terms created as you run from $n = 0$ to $n = 9$.

6. A quick check of the denominator lets you see you have a linear factor and an irreducible quadratic factor (i.e. it can't be further factorised).

This decides the strategy. $\dfrac{2x^2 + 2x + 5}{(2x + 3)(x^2 + 1)} = \dfrac{A}{2x + 3} + \dfrac{Bx + C}{x^2 + 1}$

Multiplying throughout by the denominator of the function:

$$2x^2 + 2x + 5 = A(x^2 + 1) + (Bx + C)(2x + 3)$$

6. **(continued)**

Now pick suitable values of x to set up a system of equations to solve for A, B and C.

When $x = 0$:	$5 = A + 3C$	...	①
When $x = -1$:	$5 = 2A - B + C$	...	②
When $x = 1$:	$9 = 2A + 5B + 5C$	...	③
② − 2.①:	$-5 = -B - 5C$	...	④
③ − 2.①:	$-1 = 5B - C$	...	⑤
⑤ + 5.④:	$-26 = -26C$	...	⑥

$$\Rightarrow C = 1$$

Substitute in ④: $-5 = -B - 5 \Rightarrow B = 0$

Substitute in ①: $5 = A + 3 \Rightarrow A = 2$

Thus $\dfrac{2x^2 + 2x + 5}{(2x + 3)(x^2 + 1)} = \dfrac{2}{2x + 3} + \dfrac{1}{x^2 + 1}$

$$\int \frac{2x^2 + 2x + 5}{(2x + 3)(x^2 + 1)}\, dx = \int \frac{2}{2x + 3}\, dx + \int \frac{1}{x^2 + 1}\, dx$$

$$= \ln|2x + 3| + \tan^{-1} x + C$$

Marks

1 for stating the correct form (in terms of A, B and C or equivalent);

3 for finding the values of A, B and C (lose 1 per error);

1 for logarithmic term;

1 for inverse tan term.

NOTES

Breaking down rational functions by means of the partial fractions algorithms will often produce standard forms for ln and $\tan^{-1}$ functions.

There is no set way of finding the values of the constants. I went for the 'line of least resistance'. I could have tried $x = -\dfrac{3}{2}$ which would have given me the value of A directly. However it would have involved manipulating negative numbers and fractions.

7. *(a)* $\dfrac{dy}{d\theta} = -2\sin\theta$

$$\frac{dx}{d\theta} = 2\cos 2\theta . 2 = 4\cos 2\theta$$

$$\Rightarrow \frac{dy}{dx} = \frac{\dfrac{dy}{d\theta}}{\dfrac{dx}{d\theta}} = \frac{-2\sin\theta}{4\cos 2\theta} = \frac{-\sin\theta}{2\cos 2\theta}$$

Marks

1 for finding $\dfrac{dy}{d\theta}$ and $\dfrac{dx}{d\theta}$;

1 for finding $\dfrac{dy}{dx}$.

NOTES Your working should reflect the fact that you understand the relation between the derivatives.

7. (continued)

7. *(b)* Find the value of the parameter when $y = 3$

$$y = 3 \Rightarrow 3 + 2\cos\theta = 3 \Rightarrow \cos\theta = 0 \Rightarrow \theta = \frac{\pi}{2} \text{ or } \frac{3\pi}{2}$$

Marks   **1** for finding θ when $y = 3$.

> **NOTES** This is testing your understanding of what is meant by parametric equations.

7. *(c)* Use this to find the value of x:

$x = 4 + 2\sin(2\theta)$:

When $\theta = \frac{\pi}{2}$, $x = 4 + 2\sin\frac{\pi}{2} = 4$

When $\theta = \frac{3\pi}{2}$, $x = 4 + 2\sin\frac{3\pi}{2} = 4$

Find the gradient at these instances

When $\theta = \frac{\pi}{2}$, $\dfrac{dy}{dx} = \dfrac{-\sin\left(\frac{\pi}{2}\right)}{2\cos\pi} = \dfrac{-1}{-2} = \dfrac{1}{2}$

When $\theta = \frac{3\pi}{2}$, $\dfrac{dy}{dx} = \dfrac{-\sin\left(\frac{3\pi}{2}\right)}{2\cos 3\pi} = \dfrac{1}{-2} = -\dfrac{1}{2}$

So the equations of the tangents are

When $\theta = \frac{\pi}{2}$: $y - 3 = \dfrac{1}{2}(x - 4)$ i.e. $x - 2y + 2 = 0$

When $\theta = \frac{3\pi}{2}$: $y - 3 = -\dfrac{1}{2}(x - 4)$ i.e. $x + 2y - 10 = 0$

Marks   **1** for establishing the value of x and one gradient.

1 for establishing equations (both)

> **NOTES** A diagram has been provided and it does *look* like $x = 4$ when $y = 3$ but you still have to prove it.

8. *(a)* $z = 3 + 4i$ is a root implies $z = 3 - 4i$ is another.

Marks   **1** for knowing this point of theory.

> **NOTES** If a root is $a + bi$ then a second root is $a - bi$.

8. *(b)* This means $z - (3 + 4i)$ and $z - (3 - 4i)$ are factors of the expression on L.H.S. of the equation.

This means the product $(z - (3 + 4i))(z - (3 - 4i))$ is a factor.

i.e. expanding this: $z^2 - 6z + 25$ is a factor.

We can find the complementary factor by division.

$$
\begin{array}{r}
z^2 - \ \ 2z + 2 \\
z^2 - 6z + 25 \ \big)\ \overline{z^4 - 8z^3 + 39z^2 - 62z + 50} \\
\underline{z^4 - 6z^3 + 25z^2} \\
-2z^3 + 14z^2 - 62z + 50 \\
\underline{-2z^3 + 12z^2 - 50z} \\
2z^2 - 12z + 50 \\
2z^2 - 12z + 50 \\
\underline{} \\
\end{array}
$$

8. (continued)

Zero remainder confirms that $z^2 - 6z + 25$ and $z^2 - 2z + 2$ are factors.

Solving the equation:

$$z^4 - 8z^3 + 39z^2 - 62z + 50 = 0$$

$$\Rightarrow (z^2 - 6z + 25)(z^2 - 2z + 2) = 0$$

$$\Rightarrow z^2 - 6z + 25 = 0 \text{ or } z^2 - 2z + 2 = 0$$

$z^2 - 6z + 25 = 0$ gives $z = 3 + 4i$ and $z = 3 - 4i$ as roots

$z^2 - 2z + 2 = 0$ gives $z = \dfrac{-(-2) \pm \sqrt{(-2)^2 - 4.1.2}}{2.1} = \dfrac{2 \pm \sqrt{-4}}{2} = 1 \pm i.$

Marks

1 for identifying two factors and hence quadratic factor

1 for method of finding the complementary quadratic factor (division)

1 for finding the complementary quadratic factor

1 for solving to find the four complex roots.

NOTES

- If a root is $a + bi$ then a second root is $a - bi$.
- If $a + bi$ is a root then $z - (a + bi)$ is a factor.
- When multiplying $(z - (a + bi))$ and $(z - (a - bi))$ it cuts out work to consider them as $((z - a) - bi)$ and $((z - a) + bi)$. The difference of squares can be exploited and the result is $(z - a)^2 - (bi)^2 = z^2 - 2az + a^2 + b^2$
- If you know one factor then division will produce another. As a self-check the remainder of the division will be zero.

9. (a)

$$f(x) = \frac{x^3}{x^2 - 4}$$

$$\Rightarrow f(-x) = \frac{(-x)^3}{(-x)^2 - 4} = \frac{-x^3}{x^2 - 4} = -f(x)$$

Since $f(-x) = -f(x)$ then by definition, $f(x)$ is an odd function.

Marks

1 for substituting $-x$ for x and performing some manipulaton;

1 for demonstrating $f(-x) = -f(x)$ and drawing conclusion.

NOTES

The definition of an odd function ... $f(-x) = -f(x)$.

It has half-turn symmetry about the origin.

This should help you cut down on the work in deducing features.

[c.f. The definition of an even function ... $f(-x) = f(x)$. It has bilateral symmetry about the y-axis]

9. (b)

The vertical asymptotes can be found by finding values of x which make the denominator zero.

$$x^2 - 4 = 0 \Rightarrow x = 2 \text{ or } x = -2$$

The equations of the vertical asymptotes are $x = 2$ and $x = -2$

9. (continued)

Marks   **1** for identifying vertical asymptotes.

> **NOTES** Normally the domain would have been stated in the question, viz. $x \neq \pm 2$ but that lets you know the vertical asymptotes by default.

9. (*c*) We find non-vertical asymptotes by considering how the function behaves as $x \to \infty$. Performing a division we get

$$
\begin{array}{r}
x \\
x^2 - 4 \overline{)\, x^3 } \\
\underline{x^3 - 4x} \\
4x
\end{array}
$$

Thus $f(x) = \dfrac{x^3}{x^2 - 4} = x + \dfrac{4x}{x^2 - 4} = x + \dfrac{\dfrac{4}{x}}{1 - \dfrac{4}{x^2}}$

As $x \to \infty$ we see the fractional part tends towards zero and that $f(x) \to x$.

The non-vertical asymptote is $y = x$.

Marks   **1** for expressing the function in terms of a linear function and a rational function with a numerator of lower order than the denominator.

   1 for dealing with rational part to justify that it tends to zero AND for stating the equation of the asymptote.

> **NOTES** In order to check the behaviour of a rational function at infinity, you must first make sure that the numerator is of a lower order than the numerator (by division if necessary).
> You must then divide top and bottom of the fractional part by the highest power of x.

9. (*d*) The curve cuts the asymptote $y = x$ when $x + \dfrac{4x}{x^2 - 4} = x$

$\Rightarrow \dfrac{4x}{x^2 - 4} = 0$

$\Rightarrow x = 0$

$f(0) = 0$.

The curve cuts its own non-vertical asymptote at $(0, 0)$.

Marks   **1** for forming condition for intersection

   1 for finding $(0, 0)$

> **NOTES** It is often forgotten that a curve can cut its own non-vertical asymptote.

9. (*e*) $f'(x) = \dfrac{(x^2 - 4).3x^2 - x^3.2x}{(x^2 - 4)^2} = \dfrac{x^2(x^2 - 12)}{(x^2 - 4)^2}$

At stationary points $f'(x) = 0$

$\Rightarrow$ at S.P.s $x^2(x^2 - 12) = 0$

$\Rightarrow x = 0$ or $x = \pm \sqrt{12}$

9. (continued)

x	$\rightarrow$	$-\sqrt{12}$	$\rightarrow$	$\rightarrow$	0	$\rightarrow$	$\rightarrow$	$\sqrt{12}$	$\rightarrow$
x^2	+	+	+	+	0	+	+	+	+
$(x^2-4)^2$	+	+	+	+	+	+	+	+	+
x^2-12	+	0	−	−	−	−	−	0	+
dy/dx	+	0	−	−	0	−	−	0	+
dirn	/	−	\	\	−	\	\	−	/
nature		max T.P.			horiz P.I.			min T.P.	

When $x = -\sqrt{12}$ there is a maximum turning point;

When $x = 0$ there is a horizontal point of inflexion;

When $x = \sqrt{12}$ there is a minimum turning point.

Marks   **1** for finding the derivative;

1 for finding where the derivative equals zero;

1 for determining the behaviour of the function in the neighbourhood of the S.P.s

9. (f)

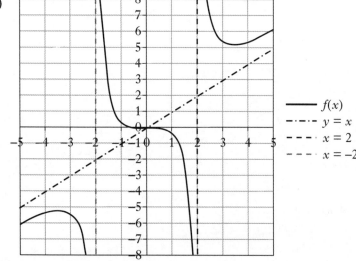

Marks   **1** for a consistent sketch.

10. (a) Separating the variables

$$\frac{dN}{dt} = k(500 - N)$$

$$\Rightarrow \int \frac{dN}{500 - N} = \int k \, dt$$

$$\Rightarrow -\ln(500 - N) = kt + C \qquad \ldots \qquad \text{①}$$

$$\Rightarrow (500 - N) = e^{-kt - C}$$

$$\Rightarrow N = 500 - e^{-kt - C}$$

Marks   1 for separating the variables

1 for performing the integration including the constant of integration

1 for making N the subject of the formula.

> **NOTES**
>
> It can be worthwhile labeling the *implicit* relation between N and t for the sake of future reference.
>
> As you will see, it can save a lot of time and effort when looking for particular values of C, k, and t for a known value of N.

10. (b) Using the result ①, $N(0) = 10 \Rightarrow -\ln(500 - 10) = k.0 + C$
$\Rightarrow C = -\ln(490)$

Marks   1 for finding expression for C consistent with earlier work.

> **NOTES**   Here is one place where using the above labelled form will save work.

10. (c) Using the result ①, $N(1) = 20 \Rightarrow -\ln(500 - 20) = k.1 - \ln(490)$
$\Rightarrow k = \ln(490) - \ln(480)$

$$\Rightarrow k = \ln\left(\frac{49}{48}\right)$$

Thus

$$N = 500 - e^{-t\ln\left(\frac{49}{48}\right) + \ln(490)}$$

$$\Rightarrow N = 500 - e^{\ln\left(\left(\frac{49}{48}\right)^{-t}\right)}.e^{\ln(490)}$$

$$\Rightarrow N = 500 - \left(\left(\frac{49}{48}\right)^{-t}\right) \cdot 490$$

$$\Rightarrow N = 500 - 490\left(\frac{48}{49}\right)^{t}$$

Marks   1 for finding N in terms of t.

> **NOTES**
>
> This looks a lot for 1 mark. However one can use ones calculator to reduce/remove the need for laws of logs from the beginning, reducing it to a 'number crunching' exercise. *Caveat:* the numerical approach could generate pretty big rounding errors by the time you get to the end of the question, though these will not be penalised.

10. (continued)

10. (*d*) When looking for *t* it is easier to use ①. When is $N = 475$?

$$-\ln(500 - 475) = \ln\left(\frac{49}{48}\right)t - \ln(490)$$

$$\Rightarrow -\ln(25) + \ln(490) = \ln\left(\frac{49}{48}\right)t$$

$$\Rightarrow \ln\left(\frac{490}{25}\right) \div \ln\left(\frac{49}{48}\right) = t$$

thus $t = 144\cdot3 \ldots$

On the 145th day when buying the 145th packet he will only have 25 stickers to collect.

Marks   **1** for substituting 475 for *N* and starting to solve for *t*.

1 for solving for *t* and interpreting answer in the context of the question.

> NOTES When you want for 25, there must be 475 stickers in your collection.

11. (*a*) By repeated division we get

$4785 = 1.3255 + 1530$	…	①
$3255 = 2.1530 + 195$	…	②
$1530 = 7.195 + 165$	…	③
$195 = 1.165 + 30$	…	④
$165 = 5.30 + 15$	…	⑤
$30 = 2.15 + 0$		

15 is the last non-zero remainder

So 15 is the greatest common divisor of 4785 and 3255

Marks   **1** for performing 3 cycles of the algorithm.

1 for completing the algorithm and interpreting result, stating the required GCD.

> NOTES So far the SQA have only ever asked the candidate to find *x* and *y* such that $ax + by = c$ where *c* is the GCD of *a* and *b*.
>
> This variation is in the syllabus and could be asked.

11. (*b*) From ⑤ you get $15 = 165 - 5.30$

From ④ you get an expression for 30: $15 = 165 - 5.(195 - 1.165) = 6.165 - 5.195$

From ③, an expression for 165: $15 = 6.(1530 - 7.195) - 5.195 = 6.1530 - 47.195$

From ②, one for 195: $15 = 6.1530 - 47.(3255 - 2.1530) = 100.1530 - 47.3255$

From ①, one for 1530: $15 = 100.(4785 - 1.3255) - 47.3255 = 100.4785 - 147.3255$

So $100.4785 - 147.3255 = 15$

If you multiply throughout by 24 (because $360 \div 15 = 24$)

you get $2400.4785 - 3528.3255 = 360$

11. (continued)

Thus two possible integers are

$x = -3528$ and $y = 2400$

Marks   1 for reversing process to express the GCD in terms of the original two numbers.

1 for multiplying by the necessary factor to get the required value.

1 for interpreting your working and stating the solution.

NOTES

You have been asked for any pair of numbers which fit the description. You may be asked for specific solutions, e.g. the smallest positive answers or the solutions with the smallest magnitude.

When you have found one solution to $ax + by = c$, viz (x, y) you can find as many more as you like

... they all take the form $(x + kb, y - ka)$ where k is any number. Though if it is a fraction, its denominator has to be a factor of c.

e.g. if (x, y) is a solution in our problem then so is $(x + 319, y - 217)$.

I've used $k = \dfrac{1}{15}$. So chipping away you can find that as well as $(-3528, 2400)$ being a solution, $(-19, 13)$ is another solution with smaller values.

12.

The equation is of the form where multiplying by an integrating factor, I, will work.

$$I = e^{\int 3x^2 dx} = e^{x^3}$$

Multiplying through by I you get

$$e^{x^3}\frac{dy}{dx} + 3x^2 e^{x^3} y = (1 + 3x^2)e^x e^{x^3}$$

By inspection this becomes

$$\frac{d}{dx}\left(e^{x^3} y\right) = (1 + 3x^2)e^{x + x^3}$$

Integrating both sides you get

$$e^{x^3} y = \int (1 + 3x^2)e^{x + x^3} dx$$

The integral can be performed using the substitution

$u = x + x^3$ so that $du = (1 + 3x^2)dx$

$$e^{x^3} y = \int e^u du$$

$$\Rightarrow e^{x^3} y = e^u + C$$

$$\Rightarrow e^{x^3} y = e^{x+x^3} + C$$

$$\Rightarrow y = e^x + \frac{C}{e^{x^3}}$$

Given $x = 0$ when $y = 1$

then $1 = 1 + C \Rightarrow C = 0$

The particular solution is $y = e^x$.

12. (continued)

Marks   **1** for finding the integrating factor, I.

1 for multiplying by I and identifying LHS with $\dfrac{d}{dx}(Iy)$

1 for using a suitable substitution to perform integration

1 for finding general solution

1 for using initial conditions to find the particular solution.

> **NOTES**
>
> When a differential equation is of the form $\dfrac{dy}{dx} + P(x)y = Q(x)$
>
> it can be solved by multiplying throughout by $I = e^{\int P(x)\,dx}$
>
> This has the effect of making the equation $\dfrac{d}{dx}(Iy) = IQ(x)$
>
> If $IQ(x)$ can be integrated, the equation can be solved.

13. (a) A matrix, A, is singular if it has no inverse. This will occur if $\det(A) = 0$.

$$\det(A) = 4\begin{vmatrix} -1 & 2 \\ 0 & y \end{vmatrix} - 1\begin{vmatrix} -5 & 2 \\ -2 & y \end{vmatrix} + x\begin{vmatrix} -5 & -1 \\ -2 & 0 \end{vmatrix}$$

$$= 4(-y - 0) - (-5y + 4) + x(0 - 2)$$

$$= -4y + 5y - 4 - 2x = y - 2x - 4$$

A is singular when $y - 2x - 4 = 0$

Marks   **1** for interpreting the question and finding det(A).

1 for finding required condition.

> **NOTES**   Vocabulary may change and singular, non-singular, invertible and non-invertible should be understood.

13. (b) $A.A^{-1} = I$

$$\begin{pmatrix} 4 & 1 & x \\ -5 & -1 & 2 \\ -2 & 0 & y \end{pmatrix} \cdot \begin{pmatrix} -3 & -3 & 1 \\ 11 & 10 & -3 \\ -2 & -2 & 1 \end{pmatrix} = \begin{pmatrix} 1 & 0 & 0 \\ 0 & 1 & 0 \\ 0 & 0 & 1 \end{pmatrix}$$

Consider the entry in the 1st row and 1st column of I.

$4.(-3) + 1.11 + x.(-2) = 1$

$\Rightarrow -12 + 11 - 2x = 1$

$\Rightarrow x = -1$

Consider the entry in the 3rd row and 1st column of I.

$-2.(-3) + 0.11 + y(-2) = 0$

$\Rightarrow 6 + 0 - 2y = 0$

$\Rightarrow y = 3$

13. (continued)

Marks   **1** for showing knowledge of matrix multiplication.

1 for equating relevant entries to find x.

1 for equating relevant entries to find y.

> **NOTES**
>
> When matrices are equal, corresponding entries are equal.
>
> This truth is the key to many questions.

14. First find the complementary function (C.F.), the solution to

$$\frac{d^2y}{dx^2} + 7\frac{dy}{dx} + 12y = 0.$$

The auxiliary equation is $m^2 + 7m + 12 = 0$

$(m + 3)(m + 4) = 0$

$\Rightarrow m = -3$ or $m = -4$

$\Rightarrow$ C.F. $= Ae^{-3x} + Be^{-4x}$

Now find a particular integral (P.I.). Since the RHS $= 24x^2 + 4x$ you should try

a quadratic … if $y = ax^2 + bx + c$ then $\frac{dy}{dx} = 2ax + b$ and $\frac{d^2y}{dx^2} = 2a$

Substituting these into the original equation:

$2a + 7(2ax + b) + 12(ax^2 + bx + c) = 24x^2 + 4x$

$\Rightarrow 12ax^2 + (14a + 12b)x + (2a + 7b + 12c) = 24x^2 + 4x$

Equating coefficients:

x^2: $12a = 24 \Rightarrow a = 2$

x: $14a + 12b = 4 \Rightarrow 28 + 12b = 4 \Rightarrow b = -2$

constant: $2a + 7b + 12c = 0 \Rightarrow 4 - 14 + 12c = 0 \Rightarrow c = \dfrac{5}{6}$

General solution: $y = $ CF + PI

i.e. $y = Ae^{-3x} + Be^{-4x} + 2x^2 - 2x + \dfrac{5}{6}$

Marks   **1** for forming the auxiliary equation

1 for solving the auxiliary equation

1 for stating the corresponding Complementary Function.

1 for choosing form of PI, finding the relevant derivatives and substituting into equation

1 for equating coefficients, finding a, b and c, and giving the general solution.

> **NOTES**
>
> - When you solve the auxiliary equation there are three basic outcomes
>
>   (i) two distinct roots (ii) two coincident roots (iii) two complex roots.
>
>   Make sure you know how to form the CF for each case.
>
> - Use the RHS to guide you as to the general form of the PI.
>
>   Guard against the PI being 'absorbed' by the CF. Know how to react in these cases.

15. (*a*) Labelling the points A(–1, 2, 4), B(1, 1, –4) and C(–2, 3, 7)

We need two vectors on the plane: $\overrightarrow{AB} = \begin{pmatrix} 1+1 \\ 1-2 \\ -4-4 \end{pmatrix} = \begin{pmatrix} 2 \\ -1 \\ -8 \end{pmatrix}$ $\overrightarrow{AC} = \begin{pmatrix} -2+1 \\ 3-2 \\ 7-4 \end{pmatrix} = \begin{pmatrix} -1 \\ 1 \\ 3 \end{pmatrix}$

Finding their vector product will give us a normal to the plane.

$$\begin{pmatrix} 2 \\ -1 \\ -8 \end{pmatrix} \times \begin{pmatrix} -1 \\ 1 \\ 3 \end{pmatrix} = \begin{vmatrix} i & j & k \\ 2 & -1 & -8 \\ -1 & 1 & 3 \end{vmatrix}$$

$$= i \begin{vmatrix} -1 & -8 \\ 1 & 3 \end{vmatrix} - j \begin{vmatrix} 2 & -8 \\ -1 & 3 \end{vmatrix} + k \begin{vmatrix} 2 & -1 \\ -1 & 1 \end{vmatrix}$$

$$= 5i + 2j + k$$

The scalar product $\begin{pmatrix} 5 \\ 2 \\ 1 \end{pmatrix} \cdot \begin{pmatrix} x \\ y \\ z \end{pmatrix} = k$ where (x, y, z) is the typical point on the plane.

We know A(–1, 2, 4) is on the plane: $\begin{pmatrix} 5 \\ 2 \\ 1 \end{pmatrix} \cdot \begin{pmatrix} -1 \\ 2 \\ 4 \end{pmatrix} = k = -5 + 4 + 4 = 3$

So the equation of the plane is $\begin{pmatrix} 5 \\ 2 \\ 1 \end{pmatrix} \cdot \begin{pmatrix} x \\ y \\ z \end{pmatrix} = 3$ or $5x + 2y + z = 3$

Marks

1 for finding two vectors on the plane

1 for finding a normal to the plane

1 for general form of the equation

1 for particular value of k and final equation.

> NOTES
>
> *The equation of a plane*: If the normal to a plane is $ai + bj + ck$, where i, j, k are the basis unit vectors, and $xi + yj + zk$ is the position vector of the typical point on the plane, their scalar product will be a constant. i.e. $ax + by + cz = t$.
>
> Find the value of this constant, t, for a particular point and you have found the equation of the plane.
>
> For our purposes, finding the *cross* product is a good way of finding a vector normal to two others.

15. (*b*) Where $2x + y + 3z = 1$ and $5x + 2y + z = 3$ intersect a line is formed.

To find the equation of a line you need (i) a point on it (ii) its direction.

(i) Choose the point where the line cuts the *x*-*y* plane i.e. $z = 0$.

$$2x + y = 1 \qquad \dots \qquad ①$$
$$5x + 2y = 3 \qquad \dots \qquad ②$$
$$2 \cdot ①: \quad 4x + 2y = 2 \qquad \dots \qquad ③$$
$$② - ③: \qquad x = 1$$

Substitute in ①: $y = -1$

The line passes through T(1, –1, 0)

15. (continued)

(ii) The line lies on both planes so it is at right angles to both normals … find the vector product of the two normals … it will be parallel to the line.

$$\begin{pmatrix} 2 \\ 1 \\ 3 \end{pmatrix} \times \begin{pmatrix} 5 \\ 2 \\ 1 \end{pmatrix} = \begin{vmatrix} i & j & k \\ 2 & 1 & 3 \\ 5 & 2 & 1 \end{vmatrix}$$

$$= i \begin{vmatrix} 1 & 3 \\ 2 & 1 \end{vmatrix} - j \begin{vmatrix} 2 & 3 \\ 5 & 1 \end{vmatrix} + k \begin{vmatrix} 2 & 1 \\ 5 & 2 \end{vmatrix} = -5i + 13j - k$$

If $P(x, y, z)$ is a typical point on the line then $\overrightarrow{TP}$ is parallel to this vector product.

The equation of the required line is:

$$\overrightarrow{TP} = t \begin{pmatrix} -5 \\ 13 \\ -1 \end{pmatrix}$$

$$\begin{pmatrix} x - 1 \\ y + 1 \\ z - 0 \end{pmatrix} = t \begin{pmatrix} -5 \\ 13 \\ -1 \end{pmatrix}$$

where t is a constant.

Equating entries

$x - 1 = -5t;\ y + 1 = 13t;\ z = -t$

You can either now go for the parametric form:

$x = -5t + 1;\ y = 13t - 1;\ z = -t$

or the symmetric form:

$$\frac{x - 1}{-5} = \frac{y + 1}{13} = \frac{z}{-1} = t$$

Marks

1 for identifying one point on the line

1 for identifying the direction of the line (a parallel vector)

1 for equation of the line of intersection in any form.

NOTES

The equation of a line: If a point is known on a line (a, b, c) and the typical point on the line is (x, y, z) then the vector $\mathbf{u} = (x - a)\mathbf{i} + (y - b)\mathbf{j} + (z - c)\mathbf{k}$ can be considered the typical vector on the line.

If a vector $\mathbf{v} = p\mathbf{i} + q\mathbf{j} + r\mathbf{k}$ is known which is parallel to the line then you know from Higher Maths that $\mathbf{u} = \mathbf{v}t$ where t is a constant.

This, in essence, is the equation of the line.

$(x - a)\mathbf{i} + (y - b)\mathbf{j} + (z - c)\mathbf{k} = pt\mathbf{i} + qt\mathbf{j} + rt\mathbf{k}$

It can be expressed in **parametric form** by equating coefficients and solving for x, y and z:

$x = a + pt;\ y = b + qt;\ z = c + rt$

It can be expressed in **symmetric form** by equating coefficients and solving for t

$$\frac{x - a}{p} = \frac{y - b}{q} = \frac{z - c}{r} = t$$

15. (continued)

15. (c) Angle between the planes is the same as the angle between the normals to the

planes, $\begin{pmatrix} 2 \\ 1 \\ 3 \end{pmatrix}$ and $\begin{pmatrix} 5 \\ 2 \\ 1 \end{pmatrix}$

$$\cos\theta = \frac{\begin{pmatrix} 2 \\ 1 \\ 3 \end{pmatrix} \cdot \begin{pmatrix} 5 \\ 2 \\ 1 \end{pmatrix}}{\left|\begin{matrix} 2 \\ 1 \\ 3 \end{matrix}\right| \times \left|\begin{matrix} 5 \\ 2 \\ 1 \end{matrix}\right|} = \frac{10+2+3}{\sqrt{14}\sqrt{30}} = 0 \cdot 73192505 \ldots$$

$\Rightarrow$ the required angle is $43 \cdot 0°$ to 1 d.p.

Marks **1** for identifying both normals

1 for knowing to use the scalar product and finding magnitudes and scalar product.

1 for finding the required angle.

> NOTES "Angle between the planes is the same as the angle between the normals to the planes" is an important point and should be stated.

16. (a) Let u_n be the nth term of the first sequence and v_n the nth term of the second.

Given $u_1 = 6$ and $S_6 = 1176$ and using the formula $S_n = \frac{n}{2}\big(2a + (n-1)d\big)$

$1176 = \frac{21}{2}(2.6 + (21-1)d)$

$\Rightarrow 1176 = 21(6 + 10d)$

$\Rightarrow d = 5$

$\Rightarrow u_2 = u_1 + 5 = 6 + 5 = 11$

Marks **1** for using relevant formula to find an expression in d alone.

1 for solving for d and finding the two terms required.

> NOTES The formula for S_n will not be given; you must remember it or know how to get it.

16. (b) $1176 = \frac{21}{2}(2.a + (21-1)d)$

$\Rightarrow a = 56 - 10d$

Marks **1** for using relevant formula to find an expression in a and d.

1 for solving for a explicitly.

> NOTES The formula is not stated again but the use of basic substitution with no calculation makes it plain what is being used.

16. (continued)

16. (c) Given $v_9 = 50$ and knowing the formula $v_n = a + (n-1)d$

$50 = 56 - 10d + 8d = 56 - 2d$

$\Rightarrow d = 3$

$\Rightarrow v_1 = 56 - 10.3 = 26$

$\Rightarrow v_2 = 29$

Marks   **1** for making relevant use of the information given.

1 for making use of the identity established in part (b)

1 for finding the required terms.

> **NOTES**   Again the formula for the *n*th term is not given in the exam.

16. (d) $v_n = u_n$

$\Rightarrow v_1 + (n-1)d_v = u_1 + (n-1)d_v$

$\Rightarrow 26 + (n-1)3 = 6 + (n-1)5$

$\Rightarrow 20 = 2(n-1)$

$\Rightarrow n = 11$

The common term is the 11th term

which is $26 + (11-1).3 = 56$

[Check: $6 + (11-1).5 = 56$]

Marks   **1** for establishing an equation relating the sequences … and involving *n*.

1 for substituting all relevant data and an equation in *n* alone.

1 for solving for *n* and finding the corresponding term.

> **NOTES**   Questions of this sort rely heavily on you remembering the formulae associated with Arithmetic Sequences:
>
> $u_n = a + (n-1)d$
>
> $S_n = \frac{n}{2}\left(2a + (n-1)d\right)$

Answers to Practice Exam 2

SOLUTIONS, NOTES AND MARKING SCHEME FOR EXAM 2

1. (a) $\left(x + \dfrac{1}{x}\right)^4 = x^4 + 4.x^3.\dfrac{1}{x} + 6.x^2.\dfrac{1}{x^2} + 4.x.\dfrac{1}{x^3} + \dfrac{1}{x^4}$

$$= x^4 + 4x^2 + 6 + 4x^{-2} + x^{-4}$$

Marks   1 for method (using binomial theorem);

1 for method (simplification).

> **NOTES**   The question is testing your knowledge of both binomial expansion and the algebraic laws of indices. Try to show by your intermediate working that you do indeed understand these facts.

1. (b) $(10.1)^4 = \left(10 + \dfrac{1}{10}\right)^4 = 10^4 + 4.10^2 + 6 + 4.10^{-2} + 10^{-4}$

$$= 10\,000 + 400 + 6 + 0.04 + 0.0001.$$

$$= 10\,406 \text{ to the nearest whole number.}$$

Marks   1 for insight to relate problems (a) and (b).

1 for method (association with part (a));

1 for processing …

> **NOTES**   The word 'Hence' does not *allow* you to get the answer by any other means. So the answer simply appearing will not receive the marks since this can be achieved by calculator alone.

2. (a) $\dfrac{d(\sec^2 x)}{dx} = \dfrac{d(\cos x)^{-2}}{dx} = -2(\cos x)^{-3}.-\sin x$

$$= 2\sec^3 x \sin x$$

… or equivalent.

Marks   1 for method (preparation of expression for differentiation);

1 for processing … performing the chain rule.

> **NOTES**   The understanding of inverse trigonometric ratios and of the chain rule are being examined. Use your working to make it clear you know both.

2. (b) $y = 3^{2x}$

$$\Rightarrow \ln y = 2x \ln 3$$

Differentiating you get

$$\frac{1}{y}\frac{dy}{dx} = 2\ln 3$$

$$\Rightarrow \frac{dy}{dx} = y.2\ln 3$$

$$\Rightarrow \frac{dy}{dx} = 2\ln 3.3^{2x}$$

2. (continued)

Marks **1** for method (using the laws of logs to the point where you are ready to perform implicit differentiation);

1 for processing ... performing implicit differentiation.

1 for expressing $\dfrac{dy}{dx}$ in terms of x only.

> **NOTES**
>
> Make sure you know the method known as logarithmic differentiation?
>
> (i)   take logs of both sides
>
> (ii) perform implicit differentiation
>
> (iii) make $\dfrac{dy}{dx}$ the subject of the resulting equation.
>
> (iv) substitute for y.

3. (a) Using $(-1, 1)$ on circle gives: $(-1)^2 + (1)^2 + 2.g.(-1) + 2.f.(1) + c = 0$

Using $(7, -3)$ on circle gives: $(7)^2 + (-3)^2 + 2.g.(7) + 2.f.(-3) + c = 0$

Using $(0, 4)$ on circle gives: $(0)^2 + (4)^2 + 2.g.(0) + 2.f.(4) + c = 0$

When simplified this leads to the system of equations:

$$
\begin{aligned}
-2g \;\; +2f \;\; + c \;&= -2 \\
14g \;\; -6f \;\; + c \;&= -58 \\
8f \;\; + c \;&= -16
\end{aligned}
$$

Marks **1** for forming the system of equations

3. (b) $\begin{pmatrix} -2 & 2 & 1 & -2 \\ 14 & -6 & 1 & -58 \\ 0 & 8 & 1 & -16 \end{pmatrix}$

$\begin{array}{l} R1 \to R1/(-2) \\ \\ R2 \to R2 - 14R1 \end{array} \begin{pmatrix} 1 & -1 & -\dfrac{1}{2} & 1 \\ 0 & 8 & 8 & -72 \\ 0 & 8 & 1 & -16 \end{pmatrix}$

$\begin{array}{l} \\ R2 \to R2/8 \\ R3 \to R3 - 8R2 \end{array} \begin{pmatrix} 1 & -1 & -\dfrac{1}{2} & 1 \\ 0 & 1 & 1 & -9 \\ 0 & 0 & -7 & 56 \end{pmatrix}$

$\begin{array}{l} \\ \\ R3 \to R3/(-7) \end{array} \begin{pmatrix} 1 & -1 & -\dfrac{1}{2} & 1 \\ 0 & 1 & 1 & -9 \\ 0 & 0 & 1 & -8 \end{pmatrix}$

3. **(continued)**

Row three: $\Rightarrow c = -8$

Row two: $\Rightarrow f - 8 = -9 \Rightarrow f = -1$

Row one: $\Rightarrow g + 1 + 4 = 1 \Rightarrow g = -4$

The equation of the circle is $x^2 + y^2 - 8x - 2y - 8 = 0$

Marks 1 for method (obtaining the augmented matrix);

2 marks for manipulation to upper triangular form (lose one mark per error);

1 for value of c;

1 for values of g and f and the equation of the circle.

NOTES

Only a solution obtained by Gaussian Elimination is acceptable for marks.

The augmented matrix plus clear elementary row operations (EROs) to achieve upper triangular form are essential.

Thereafter substitution is acceptable.

Indicate what operations you have chosen to perform. This will make checking your working afterwards easier.

4. $x^2 + y^2 + xy = 9$

Differentiating gives

$$2x + 2y\frac{dy}{dx} + x\frac{dy}{dx} + y = 0$$

$$\Rightarrow \frac{dy}{dx}(2y + x) = -2x - y$$

$$\Rightarrow \frac{dy}{dx} = \frac{-2x - y}{2y + x}$$

(a) The gradient is 1 when the derivative is 1, when $-2x - y = 2y + x \Rightarrow y = -x$

The line $y = -x$ cuts the ellipse $x^2 + y^2 + xy = 9$ when

$(-y)^2 + y^2 + (-y)y = 9$

$\Rightarrow y^2 = 9$

$\Rightarrow y = \pm 3$

The required points are $(-3, 3)$ and $(3, -3)$.

(b) The gradient is 0 when the derivative is 0, when $-2x - y = 0 \Rightarrow y = -2x$

$x^2 + (-2x)^2 + x(-2x) = 9$

$\Rightarrow 3x^2 = 9$

$\Rightarrow x = \pm\sqrt{3}$

The required points are $\left(\sqrt{3}, -2\sqrt{3}\right)$ and $\left(-\sqrt{3}, 2\sqrt{3}\right)$

(c) The gradient is undefined when the denominator of the derivative is 0, when $x + 2y = 0 \Rightarrow x = -2y$

$(-2y)^2 + y^2 + (-2y)y = 9$

$\Rightarrow 3y^2 = 9$

$\Rightarrow y = \pm\sqrt{3}$

The required points are $\left(-2\sqrt{3}, \sqrt{3}\right)$ and $\left(2\sqrt{3}, -\sqrt{3}\right)$

4. (continued)

Marks
1 for finding an explicit expression for the derivative

1 for analyzing conditions for derivative to equal 1, to the point where an equation in one variable is obtained.

1 for solving this equation and finding the required points.

1 for analyzing conditions for derivative to equal 0, to the point where an equation in one variable is obtained.

1 for solving this equation and finding the required points.

1 for analysing situation where gradient is undefined leading to the required points.

NOTES
The derivative gives a formula for the gradient. Although you can estimate where the derivative is 1, 0 or undefined from the graph, it is only by an analytical approach that marks will be obtained.

5. (a) Required to prove: $x^2 + 2x - 4 > 0 \ \forall \ x > 1; \ x \in \mathbf{Z}^+$

Consider $x = 2$ (the lowest value for which the statement is to be proved true)

When $x = 2$, $x^2 + 2x - 4 = 2^2 + 2.2 - 4 = 4$

Since $4 > 0$ the statement is true for $x = 2$.

Suppose the statement is true for some value of x viz. $x = k$.

Then $k^2 + 2k - 4 > 0$

Consider the expression when $x = k + 1$

$(k + 1)^2 + 2(k + 1) - 4 = k^2 + 2k + 1 + 2k + 2 - 4$

$= (k^2 + 2k - 4) + 2k + 3$

Since $2k + 3 > 0 \ \forall k > 1; \ k \in \mathbf{Z}^+$

then, $k^2 + 2k - 4 > 0 \Rightarrow (k^2 + 2k - 4) + 2k + 3 > 0 \ \forall k > 1; \ k \in \mathbf{Z}^+$

Thus if the statement is true for $x = k$ then it is true for $x = k + 1$.

Since it *is* true for $x = 2$ then, by induction, it is true $\forall x > 1; \ x \in \mathbf{Z}^+$

Marks
1 for showing proposal true for $x = 2$ (LHS > 0 when $x = 2$);

1 for assuming true for $x = k$ and starting to consider $x = k + 1$;

1 for using the assumption correctly;

1 for establishing that true for $x = k$ implies true for $x = k + 1$;

1 for pulling it all together in a final conclusion.

NOTES
The initial question is testing your ability to perform proof by induction and also testing your understanding of this particular form. So proof by any method other than by induction will receive no credit.

- Show that the proposal is true for the lowest case

  ... in this example $x = 2$.

- Show that if it is true for $x = k$ then it is true for $x = k + 1$.

- Draw a conclusion that with the above two bullet points demonstrated as true

  then, by induction the proposal is true for all x bigger than or equal to the lowest case.

5. **(continued)**

5. (b) Completing the square:

$x^2 + 2x - 4 = (x + 1)^2 - 5$

If $x \geq 2$ then $(x + 1)^2 \geq (2 + 1)^2$

i.e. $(x + 1)^2 \geq 9$

Thus $(x + 1)^2 - 5 \geq 4$

Since $4 > 0$

then $x^2 + 2x - 4 > 0 \ \forall x > 1; \ x \in \mathbb{Z}^+$

Marks   **1** for starting a valid approach;

1 for completing a direct proof.

NOTES

The symbol $\forall$ means "for all"

The marks here are for showing what you know a direct proof is, and for carrying it through to completion.

There are other direct proofs:

e.g. $\dfrac{dy}{dx} = 2x + 2$

Thus $\dfrac{dy}{dx} > 0 \ \forall x > -1$.

i.e. $x^2 + 2x - 4$ is an increasing function $\forall x > -1$.

Since at $x = 2$ the function is positive and increasing then it's positive $\forall x \geq 2$.

A proof by contradiction will receive no credit.

6. (a) $x^2 - 2x + 10 = (x - 1)^2 + 9$

Marks   **1** for completing the square

NOTES

This is a Higher Maths skill and may be given no credit in the actual exam. Indeed, the "guidance" that part (a) gives to allow you access to part (b) may not be given. In this case you would get a mark, but it would be for spotting the strategy.

(b) If $u = x - 1$ then $du = dx$ and $x = u + 1$

So $\displaystyle\int \frac{2x + 1}{x^2 - 2x + 10}\,dx = \int \frac{2x + 1}{(x-1)^2 + 9}\,dx = \int \frac{2(u + 1) + 1}{u^2 + 9}\,dx$

$= \displaystyle\int \frac{2u + 3}{u^2 + 9}\,du = \int \frac{2u}{u^2 + 9}\,du + 3\int \frac{1}{u^2 + 9}\,du = \int \frac{2u}{u^2 + 9}\,du + 3\int \frac{1}{u^2 + 3^2}\,du$

$= \ln\left|u^2 + 9\right| + 3 \cdot \dfrac{1}{3}\tan^{-1}\left(\dfrac{u}{3}\right) + c$

$= \ln\left|(x - 1)^2 + 9\right| + \tan^{-1}\left(\dfrac{x - 1}{3}\right) + c$

Marks   **1** for using part (a);

1 for performing the substitution;

1 for manipulation into *standard forms*

1 for integration using $\displaystyle\int \frac{f'(x)}{f(x)}\,dx = \log_e[f(x)]$

1 for integration using $\dfrac{1}{a}\tan^{-1}\left(\dfrac{x}{a}\right)$ and resubstituting x.

6. (continued)

7. (a)

$$B^2 = \begin{pmatrix} \frac{1}{\sqrt{2}} & -\frac{1}{\sqrt{2}} \\ \frac{1}{\sqrt{2}} & \frac{1}{\sqrt{2}} \end{pmatrix}\begin{pmatrix} \frac{1}{\sqrt{2}} & -\frac{1}{\sqrt{2}} \\ \frac{1}{\sqrt{2}} & \frac{1}{\sqrt{2}} \end{pmatrix} = \begin{pmatrix} \frac{1}{2}-\frac{1}{2} & -\frac{1}{2}-\frac{1}{2} \\ \frac{1}{2}+\frac{1}{2} & -\frac{1}{2}+\frac{1}{2} \end{pmatrix} = \begin{pmatrix} 0 & -1 \\ 1 & 0 \end{pmatrix}$$

Transforming (x, y):

$$\begin{pmatrix} 0 & -1 \\ 1 & 0 \end{pmatrix}\begin{pmatrix} x \\ y \end{pmatrix} = \begin{pmatrix} 0 - y \\ x + 0 \end{pmatrix} = \begin{pmatrix} -y \\ x \end{pmatrix}$$

Thus $(x, y) \to (-y, x)$ which is an anti-clockwise rotation about the origin of $90°$.

(b)

$$B^4 = B^2.B^2 = \begin{pmatrix} 0 & -1 \\ 1 & 0 \end{pmatrix}.\begin{pmatrix} 0 & -1 \\ 1 & 0 \end{pmatrix} = \begin{pmatrix} -1 & 0 \\ 0 & -1 \end{pmatrix} = -1\begin{pmatrix} 1 & 0 \\ 0 & 1 \end{pmatrix} = -1I$$

$I = -B^4$

$\Rightarrow I.B^{-1} = -B^4.B^{-1}$

$\Rightarrow B^{-1} = -B^3.B.B^{-1} = -B^3.I$

$\Rightarrow B^{-1} = -B^3$

(c)

$$AB = \begin{pmatrix} \sqrt{2} & \sqrt{2} \\ -\sqrt{2} & \sqrt{2} \end{pmatrix}\begin{pmatrix} \frac{1}{\sqrt{2}} & -\frac{1}{\sqrt{2}} \\ \frac{1}{\sqrt{2}} & \frac{1}{\sqrt{2}} \end{pmatrix} = \begin{pmatrix} 1+1 & -1+1 \\ -1+1 & 1+1 \end{pmatrix} = \begin{pmatrix} 2 & 0 \\ 0 & 2 \end{pmatrix}$$

Transforming (x, y):

$$\begin{pmatrix} 2 & 0 \\ 0 & 2 \end{pmatrix}\begin{pmatrix} x \\ y \end{pmatrix} = \begin{pmatrix} 2x + 0 \\ 0 + 2y \end{pmatrix} = \begin{pmatrix} 2x \\ 2y \end{pmatrix}$$

Under this transformation $(x, y) \to (2x, 2y)$ which is a dilatation (enlargement) scale factor 2 and centre of dilatation $(0, 0)$.

Marks 　**1** for squaring a 2×2 matrix.

1 for a *justified* geometric interpretation.

1 for squaring a 2×2 matrix and identifying the identity matrix.

1 for using your findings and introducing the inverse successfully.

1 for finding the matrix product AB.

1 for a *justified* geometric interpretation.

7. (continued)

8. (a) The essence of the diagram is:

Using Similar triangles we get

$$\frac{h}{r} = \frac{200}{50} = 4$$

Thus $h = 4r$.

Now $V = \frac{1}{3}\pi r^2 h$

So, in terms of r alone, $V = \frac{4}{3}\pi r^3$

Marks   **1** for using similar triangles to relate r to h.

1 for finding the required volume in terms of r and h.

1 for final expression.

8. (b) The second sentence on the question is telling us that $\frac{dV}{dt} = -100$ ml/sec

The question is asking for $\frac{dr}{dt}$ in cm/sec

These rates are related by the equation $\frac{dr}{dt} = \frac{dr}{dV} \times \frac{dV}{dt}$

Since $V = \frac{4}{3}\pi r^3$ then $\frac{dV}{dr} = 4\pi r^2$

So $\frac{dr}{dt} = \frac{1}{4\pi r^2} \times -100 = -\frac{25}{\pi r^2}$

When $r = 20$, $\frac{dr}{dt} = -\frac{25}{\pi.400} = -0\cdot 020$ cm/sec (to 3 d.p.)

Radius is decreasing by $0\cdot 020$ cm/sec

Marks   **1** for interpreting sentence 2 to get $\frac{dV}{dt} = -100$.

1 for using (a) to find $\frac{dV}{dr}$ or $\frac{dr}{dV}$ and using 'chain rule'.

1 for using 'chain rule'.

1 for finding formula and subsequently rate of change of r when $r = 20$ cm.

8. (continued)

NOTES

You are expected to create your mathematical model, so you should 'translate' the sentences in English into equivalent mathematical expressions i.e. use mathematical notation appropriately.

You must also state how the various rates are related [rate of change of radius with time = rate of change of radius per unit volume multiplied by the rate of change of volume with time ... $\frac{dr}{dt} = \frac{dr}{dV} \times \frac{dV}{dt}$]

9.

$$\frac{dP}{dt} = \frac{t^2}{100}(100 - P)$$

Separating the variables P and t,

$$\int \frac{dP}{100 - P} = \int \frac{t^2}{100} dt$$

Perform the integrals

$$-\ln|100 - P| = \frac{t^3}{300} + c$$

Make P the subject of the formula

$$100 - P = e^{-\frac{t^3}{300} - c}$$

$$\Rightarrow P = 100 - e^{-\frac{t^3}{300} - c}$$

This can be simplified, and under normal circumstances it should be:

$$P = 100 - e^{-\frac{t^3}{300}} \cdot e^{-c} = 100 - Ae^{-\frac{t^3}{300}} \text{ where } A \text{ is a constant equal to } e^{-c}.$$

When $t = 0$, $P = 0$ so

$$P = 100 - Ae^{-\frac{t^3}{300}}$$

$$\Rightarrow 0 = 100 - Ae^0 = 100 - A$$

$$\Rightarrow A = 100$$

Thus the required formula is $P = 100 - 100e^{-\frac{t^3}{300}} = 100\left(1 - e^{-\frac{t^3}{300}}\right)$

Marks

1 for method.

1 for successful separation expressed as equated integrals.

1 for performing integrations.

1 for making P the subject of the formula.

1 for using initial conditions correctly with an aim of finding the constant of integration.

1 for finding constant and finding the required expression for P.

NOTES

Solving differential equations, where the variables are separable is a common type of question. It often crops up in the development of growth models.

10. (a) Given $(4 + 3i)(x + iy) = 1 - i$

$$\Rightarrow x + iy = \frac{1-i}{4+3i}$$

$$\Rightarrow x + iy = \frac{(1-i)(4-3i)}{(4+3i)(4-3i)} = \frac{4-3i-4i-3}{16+9}$$

$$\Rightarrow x + iy = \frac{1-7i}{25} = \frac{1}{25} - \frac{7}{25}i$$

So $x = \dfrac{1}{25}$ and $y = -\dfrac{7}{25}$

Marks **1** for method … performing division (or multiplication) … see note below

1 for employing complex conjugate correctly (or setting up system of equations)

1 for equating parts and finding x and y.

> **NOTES**
> - If $a.b = c$ then $a = c/b$
> - The denominator can be made real by multiplying numerator and denominator by the conjugate of the denominator.
> - If $a + ib = c + id$ then $a = c$ and $b = d$.
> - an alternative strategy would be to (i) multiply out (ii) equate parts (iii) form a system of equations and solve simultaneously for x and y.

10. (b) $a + ib = \sqrt{(5+12i)}$

$$\Rightarrow 5 + 12i = (a + ib)^2$$

$$\Rightarrow 5 + 12i = a^2 + 2abi + (ib)^2$$

$$\Rightarrow 5 + 12i = a^2 - b^2 + 2abi$$

Equating real parts: $a^2 - b^2 = 5$ $\qquad$ … $\qquad$ ①

Equating imaginary parts: $2ab = 12$ $\qquad$ … $\qquad$ ②

From ②, $b = \dfrac{12}{2a} = \dfrac{6}{a}$

Substituting for b in ①, $a^2 - \left(\dfrac{6}{a}\right)^2 = 5$

$$\Rightarrow a^4 - 36 = 5a^2$$

$$\Rightarrow a^4 - 5a^2 - 36 = 0$$

$$\Rightarrow (a^2 - 9)(a^2 + 4) = 0$$

$$\Rightarrow a = \pm 3 \text{ or } a = \pm 2i$$

However a is a real number, so $a = \pm 3$.

and $b = \dfrac{6}{\pm 3} = \pm 2$

Thus $a = 3$, $b = 2$ or $a = -3$, $b = -2$ are the two solutions.

Marks **1** for method … squaring

1 for employing complex conjugate correctly.

1 for solving for the two sets for a and b.

10. (continued)

> **NOTES**
> - $x = \sqrt{y} \Rightarrow x^2 = y$
> - If $a + ib = c + id$ then $a = c$ and $b = d$.
> - $a^4 - 5a^2 - 36 = 0$ can be treated as a quadratic form.

11. (a) $u_1 = 810$; $u_n = 810r^{n-1}$; $u_5 = 810r^4$

$$\Rightarrow 810r^4 = 10$$

$$\Rightarrow r = \pm\frac{1}{3}$$

In either case $-1 < r < 1$ which is the condition for a sum to infinity to exist.

Marks   **1** for showing conditions for sum to infinity exist.

> **NOTES** The formulae for summation will not be given in the exam. These must be memorised.

11. (b) $S_\infty = \dfrac{a}{1-r} = \dfrac{810}{1 \pm \dfrac{1}{3}} = 607\dfrac{1}{2}$ or $1215 \cdot$

Since you are given that this sum should be 1215 then $r = \dfrac{1}{3}$.

Hence $u_1 = 810$; $u_2 = 270$; $u_3 = 90$; $u_4 = 30$; $u_5 = 10$

Marks   **1** for applying formula for sum to infinity.

1 For deciding which value of r applies.

1 For listing the first five values of the sequence.

> **NOTES** Make sure you explore both possible values of r.

11. (c) $u_n < 0 \cdot 01$

$$\Rightarrow 810 \cdot \left(\frac{1}{3}\right)^{n-1} < 0 \cdot 01$$

$$\Rightarrow \left(\frac{1}{3}\right)^{n-1} < \frac{0 \cdot 01}{810}$$

$$\Rightarrow (n-1)\ln\left(\frac{1}{3}\right) < \ln\left(\frac{0 \cdot 01}{810}\right)$$

$$\Rightarrow n - 1 > \frac{\ln\left(\dfrac{0 \cdot 01}{810}\right)}{\ln\left(\dfrac{1}{3}\right)}$$

$$\Rightarrow n > 11 \cdot 2877\ldots$$

So the 12th term is the first one less than $0 \cdot 01$.

Marks   **1** for setting up and starting to solve inequality.

1 For solving and interpreting solution.

> **NOTES**
> - The log function is an increasing function. So if $a < b$ then $\ln(a) < \ln(b)$
> - $\ln(\frac{1}{3})$ is negative. So $a\ln(\frac{1}{3}) < b \Rightarrow a > b \div \ln(\frac{1}{3})$

12. $y = x^2\sqrt{(\ln x)} \Rightarrow y^2 = x^4 \ln x$

$$V = \pi \int_1^4 x^4 \ln x \; dx$$

Integrating by parts

$$V = \pi \left[\ln x \cdot \frac{x^5}{5} - \int \frac{x^5}{5} \cdot \frac{1}{x} dx \right]_1^4$$

$$V = \pi \left[\ln x \cdot \frac{x^5}{5} - \frac{x^5}{25} \right]_1^4$$

$$V = \pi \left[\left(\ln 4 \cdot \frac{4^5}{5} - \frac{4^5}{25} \right) - \left(\ln 1 \cdot \frac{1^5}{5} - \frac{1^5}{25} \right) \right]_1^4$$

$$V \approx 763$$

The volume is 763 cubic units to the nearest cubic unit.

Marks   **1** for applying the formula to form the appropriate integral.

1 for method of using 'by parts'.

2 for finding the integral.

1 for using the limits.

1 for evaluation and relating to context.

> NOTES   Although in this question the formula for the volume of revolution has been given, it is a required piece of knowledge and the question could appear without this hint.

13. (a) Find a vector parallel to the line: $\begin{pmatrix} 3 \\ 1 \\ 4 \end{pmatrix} - \begin{pmatrix} 1 \\ -1 \\ 3 \end{pmatrix} = \begin{pmatrix} 2 \\ 2 \\ 1 \end{pmatrix}$

Using $(1, -1, 3)$ and the typical point (x, y, z): $\begin{pmatrix} x \\ y \\ z \end{pmatrix} - \begin{pmatrix} 1 \\ -1 \\ 3 \end{pmatrix} = \begin{pmatrix} x-1 \\ y+1 \\ z-3 \end{pmatrix}$

Since these two vectors are parallel we get

$$\begin{pmatrix} x-1 \\ y+1 \\ z-3 \end{pmatrix} = t \begin{pmatrix} 2 \\ 2 \\ 1 \end{pmatrix} = \begin{pmatrix} 2t \\ 2t \\ t \end{pmatrix} \text{ where } t \text{ is a constant.}$$

Equating entries:

$x - 1 = 2t; \; y + 1 = 2t; \; z - 3 = t \ldots$ the parametric form of the equation of the line.

or $\dfrac{x-1}{2} = \dfrac{y+1}{2} = \dfrac{z-3}{1} = t$

which is the equation of the line in symmetric form.

13. (continued)

Marks   **1** for finding a vector parallel to the line in question.

1 for finding the typical vector on the line expressed in terms of x, y and z.

1 for exploiting the relationship between parallel vectors.

1 for expressing the equation of the line in the required form.

> **NOTES**   If A and B are two points then $\mathbf{b} - \mathbf{a}$ is a vector parallel to AB.
>
> If two vectors $\mathbf{a}$ and $\mathbf{b}$ are parallel then $\mathbf{a} = t\mathbf{b}$ where t is a scalar constant.

13. (b)   If the point $(-1, 2, 0)$ also lies on the plane then $\begin{pmatrix} -1 \\ 2 \\ 0 \end{pmatrix} - \begin{pmatrix} 1 \\ -1 \\ 3 \end{pmatrix} = \begin{pmatrix} -2 \\ 3 \\ -3 \end{pmatrix}$ is a vector on the plane.

A **normal to the plane** can be found by taking the vector product of two vectors on the plane e.g. $\begin{pmatrix} 2 \\ 2 \\ 1 \end{pmatrix}$ and $\begin{pmatrix} -2 \\ 3 \\ -3 \end{pmatrix}$

$$\begin{pmatrix} 2 \\ 2 \\ 1 \end{pmatrix} \times \begin{pmatrix} -2 \\ 3 \\ -3 \end{pmatrix} = \begin{vmatrix} \mathbf{i} & \mathbf{j} & \mathbf{k} \\ 2 & 2 & 1 \\ -2 & 3 & -3 \end{vmatrix} = \mathbf{i} \begin{vmatrix} 2 & 1 \\ 3 & -3 \end{vmatrix} - \mathbf{j} \begin{vmatrix} 2 & 1 \\ -2 & -3 \end{vmatrix} + \mathbf{k} \begin{vmatrix} 2 & 2 \\ -2 & 3 \end{vmatrix}$$

$$= -9\mathbf{i} + 4\mathbf{j} + 10\mathbf{k}$$

$$= \begin{pmatrix} -9 \\ 4 \\ 10 \end{pmatrix}$$

For any point P(x, y, z) on the plane, $\mathbf{n}.\mathbf{p} = k$ where $\mathbf{n}$ is a normal and k is a scalar constant.

The **equation of the plane** is of the form

$$\begin{pmatrix} -9 \\ 4 \\ 10 \end{pmatrix} \cdot \begin{pmatrix} x \\ y \\ z \end{pmatrix} = k \Rightarrow -9x + 4y + 10z = k$$

Since $(-1, 2, 0)$ lies on the plane then k $= -9. -1 + 4.2 + 10.0 = 17$

Thus the equation of the plane is $-9x + 4y + 10z = 17$

Marks   **1** for forming a second vector on the plane and forming the cross-product

1 for finding the cross-product and the general form of the plane

1 for finding the constant and the equation of the plane.

13. (continued)

13. (c) The given line is parallel to $\begin{pmatrix} 3 \\ 1 \\ 5 \end{pmatrix}$ ... obtained by considering the coefficients of t.

The normal to the plane is $\begin{pmatrix} -9 \\ 4 \\ 10 \end{pmatrix}$

The angle between these two vectors can be obtained from $\cos\theta = \dfrac{\boldsymbol{a.b}}{|\boldsymbol{a}||\boldsymbol{b}|}$

$$\left\| \begin{pmatrix} 3 \\ 1 \\ 5 \end{pmatrix} \right\| = \sqrt{3^2 + 1^2 + 5^2} = \sqrt{35} ; \quad \left\| \begin{pmatrix} -9 \\ 4 \\ 10 \end{pmatrix} \right\| = \sqrt{(-9)^2 + 4^2 + 10^2} = \sqrt{197}$$

$$\begin{pmatrix} 3 \\ 1 \\ 5 \end{pmatrix} \cdot \begin{pmatrix} -9 \\ 4 \\ 10 \end{pmatrix} = -27 + 4 + 50 = 27$$

$$\cos\theta = \frac{27}{\sqrt{35 \times 197}} = 0 \cdot 32516...$$

So the angle between the normal and the given line is $71 \cdot 0°$

So the angle between the line and the plane itself $= 90 - 71 \cdot 0 = 19 \cdot 0°$

Marks   2 for using the scalar product to find the angle between the line and a normal.;

1 for using this to find the angle between the line and the plane.

14. (a) $\sin x \approx x - \dfrac{x^3}{3!} + \dfrac{x^5}{5!} - \dfrac{x^7}{7!}$

Marks   1 for giving the right number of terms.

14. (continued)

14. (b)

$$\sin 2x \approx 2x - \frac{(2x)^3}{3!} + \frac{(2x)^5}{5!} - \frac{(2x)^7}{7!} = 2x - \frac{8}{6}x^3 + \frac{32}{120}x^5 - \frac{128}{5040}x^7$$

Marks   **1** for giving the right number of terms in the required form.

> **NOTES** The expression found for sin x is also valid when x is exchanged for any other function of x, in this case $2x$. The expression must be worked out to form a power series i.e. $a_0 x^0 + a_1 x^1 + a_2 x^2 + a_3 x^3 + a_4 x^4 + \dots$

14. (c)

$$\sin\left(2x + \frac{\pi}{3}\right) \approx 2x + \frac{\pi}{3} - \frac{\left(2x + \frac{\pi}{3}\right)^3}{3!} + \frac{\left(2x + \frac{\pi}{3}\right)^5}{5!} - \frac{\left(2x + \frac{\pi}{3}\right)^7}{7!}$$ but this is not a power series.

We should expand the expression first:

$$\sin\left(2x + \frac{\pi}{3}\right) = \sin 2x \cos \frac{\pi}{3} + \cos 2x \sin \frac{\pi}{3} = \frac{1}{2}\sin 2x + \frac{\sqrt{3}}{2}\cos 2x$$

Now $$\cos 2x \approx 1 - \frac{(2x)^2}{2!} + \frac{(2x)^4}{4!} - \frac{(2x)^6}{6!}$$

$$\frac{1}{2}\sin 2x + \frac{\sqrt{3}}{2}\cos 2x = \frac{1}{2}\left(2x - \frac{8}{6}x^3 + \frac{32}{120}x^5 - \frac{128}{5040}x^7\right) + \frac{\sqrt{3}}{2}\left(1 - \frac{4}{2}x^2 + \frac{16}{24}x^4 - \frac{64}{720}x^6\right)$$

$$\frac{1}{2}\sin 2x + \frac{\sqrt{3}}{2}\cos 2x = \frac{\sqrt{3}}{2} + x - \sqrt{3}x^2 + \frac{2}{3}x^3 + \frac{\sqrt{3}}{3}x^4 + \frac{2}{15}x^5 + \frac{2\sqrt{3}}{45}x^6 - \frac{4}{315}x^7$$

Marks   **1** for method. (expansion and substitution)

1 for expansion of $\cos 2x$.

1 for substitution of Maclaurin expansions.

1 for expressing expansion as a power series.

> **NOTES** Since a power series is requested, a simple substitution of an expression will not do.

15. (a) Find the GCD of 198 and 210 using the Euclidean algorithm.

$210 = 1.198 + 12 \quad \dots$ ①

$198 = 16.12 + 6 \quad \dots$ ②

$12 = 2.6 + 0$

The GCD of 198 and 210 is 6.

Marks   **1** for applying Euclidean Algorithm.

1 for finding the GCD of 198 and 210.

> **NOTES** This answer might well be 'spotable' but the question is providing an opportunity for testing your knowledge of the Euclidean Algorithm.

15. (continued)

15. (*b*) Express 6 in terms of 198 and 210 by stepping through the algorithm 'backwards'.

From ②, we get $6 = 1.198 - 16.12$

From ①, we get $6 = 1.198 - 16.(1.210 - 1.198) = 17.198 - 16.210$

This gives us $(17).198 + (-16).210 = 6$

Note 6 also divides 3090 $(3090 \div 6 = 515)$.

Multiply throughout by 515:

$(17.515).198 + (-16.515).210 = 6.515$

$\Rightarrow (8755).198 + (-8240).210 = 3090$

making $x = 8755$ and $y = -8240$ a solution to the equation

and $(8755, -8240)$ a point on the line with integer coordinates.

Marks   **1** for expressing the GCD in the form $198x + 210y$.

1 for relating the above to the equation $198x + 210y = 3090$

and identifying a solution to the equation $198x + 210y = 3090$.

15. (*c*) If (a, b) is a point on the line then $198a + 210b = 3090$

Does the point $(a + 210k, b - 198k)$ satisfy the equation?

$198(a + 210k) + 210(b - 198k)$

$= 198a + 51580k + 210b - 51580k$

$= 198a + 210b$

$= 3090$

So $(a + 210k, b - 198k)$ satisfies the equation.

Marks   **1** for start to proof.

1 for reaching conclusion that (a, b) satisfies the equation $\Rightarrow (a + 210k, b - 198k)$ satisfies the equation

NOTES   So far the SQA question dealing with this area of work has avoided practical examples, simply asking for the GCD of two numbers a and b, to be expressed in the form $ax + by$ where x and y are integers. However there is nothing to stop this variety of the question type being asked.

16. The equation can be written as $\dfrac{d^2y}{dx^2} - 8\dfrac{dy}{dx} + 16y = 16x - 4$

The auxiliary equation is then $m^2 - 8m + 16 = 0$

$\Rightarrow (m - 4)^2 = 0$

$\Rightarrow m = 4$ (twice)

Since you have two coincident roots, the complementary function (CF) is $y = Ae^{4x} + Bxe^{4x}$ where A and B are arbitrary constants.

Since the right-hand side of the equation is linear you can try the form $y = ax + b$

when looking for a particular integral.

16. (continued)

$y = ax + b$

$\Rightarrow \dfrac{dy}{dx} = a$

$\Rightarrow \dfrac{d^2y}{dx^2} = 0$

Substituting these into the original equation gives:

$0 - 8a + 16(ax + b) = 16x - 4$

$\Rightarrow 16ax + 16b - 8a = 16x - 4$

Equating coefficients:

$16a = 16 \Rightarrow a = 1$

$16b - 8a = -4 \Rightarrow 16b - 8 = -4 \Rightarrow b = \dfrac{1}{4}$

The particular integral (PI) is $y = x + \dfrac{1}{4}$

The *General Solution* is $y = \mathrm{CF} + \mathrm{PI}$

i.e. $y = Ae^{4x} + Bxe^{4x} + x + \dfrac{1}{4}$

Given that when $x = 0,\ y = \dfrac{5}{4}$ then $Ae^0 + B.0.e^0 + 0 + \dfrac{1}{4} = \dfrac{5}{4} \Rightarrow A = 1$.

Now $\dfrac{dy}{dx} = 4Ae^{4x} + Be^{4x} + 4Bxe^{4x} + 1$

Given that when

$x = 0, \dfrac{dy}{dx} = 0$ then $4Ae^0 + Be^0 + 4B.0.e^0 + 1 = 0 \Rightarrow 4A + B + 1 = 0$

$\Rightarrow 4 + B + 1 = 0 \Rightarrow B = -5$

Hence the particular solution is $y = e^{4x} - 5xe^{4x} + x + \dfrac{1}{4}$

Marks   **1** for forming the auxiliary equation

1 for solving the auxiliary equation

1 for stating the corresponding Complementary Function.

1 for choosing form of PI, finding the relevant derivatives and substituting into equation

1 for equating coefficients, finding a, b and c, and giving the general solution.

1 for using the initial conditions to find A and B and hence the particular solution.

NOTES

- When the auxiliary equation produced two distinct roots p and q, the CF is

  $y = Ae^{px} + Be^{qx}$

  However if $p = q$, this collapses to $(A + B)e^{px}$ which has effectively only one arbitrary constant viz. $A + B$.

  When the roots are coincident remember the CF is of the form

  $y = Ae^{4x} + Bxe^{4x}$.

- When the PI is going to produce a term which will be effectively absorbed by the CF then a similar ploy must be used. Instead of using $y = f(x)$ for the PI, the form $y = xf(x)$ should be considered.

Answers to Practice Exam 3

SOLUTIONS, NOTES AND MARKING SCHEME FOR EXAM 3

1. (a) $y = (2x + 1)\cos^2 x$

$$\Rightarrow \frac{dy}{dx} = (2x + 1).2\cos x. -\sin x + \cos^2 x.2$$

$$\Rightarrow \frac{dy}{dx} = 2\cos^2 x - 2(2x + 1)\sin x \cos x$$

Marks   **1** for method (using product rule);

            1 for method (using chain rule on the $\cos^2 x$ term).

            1 for interpretation of function and communication of result.

> **NOTES**
> The product rule and the chain rule are both being tested here.
> The working should reflect the steps taken. The use of the 'dot' for multiplication helps the clarity.

1. (b) $y = x^x$

$$\Rightarrow \ln y = \ln(x^x) = x\ln x$$

$$\Rightarrow \frac{1}{y}\frac{dy}{dx} = x.\frac{1}{x} + 1.\ln x = 1 + \ln x$$

$$\Rightarrow \frac{dy}{dx} = (1 + \ln x)y = (1 + \ln x)x^x$$

Marks   **1** for method (taking logs of both sides);

            1 for processing … implicit differentiation.

            1 for communication … expressing the derivative in terms of x alone.

> **NOTES**
> Whenever the function to be derived contains x in the exponent or contains 'lots' of products and powers, you should consider this technique, called logarithmic differentiation.

2. Set up the augmented matrix.

$$\begin{pmatrix} 1 & 1 & 1 & 2 \\ 2 & -1 & 2t-13 & -9 \\ 4 & 3 & -1 & -3 \end{pmatrix}$$

Step 1: exchange rows to make the variable element in the last row.

$$R_2 \leftrightarrow R_3 \quad \begin{pmatrix} 1 & 1 & 1 & 2 \\ 4 & 3 & -1 & -3 \\ 2 & -1 & 2t-13 & -9 \end{pmatrix}$$

2. (continued)

Step 2: Entry a_{11} is already 1 so perform the row operations to make the entries under it zero:

$$\begin{array}{c} \\ R_2 \to R_2 - 4R_1 \\ R_3 \to R_3 - 2R_1 \end{array} \begin{pmatrix} 1 & 1 & 1 & 2 \\ 0 & -1 & -5 & -11 \\ 0 & -3 & 2t-15 & -13 \end{pmatrix}$$

Step 3: Make entry $a_{22} = 1$ and then perform the ERO to make the element under it zero:

$$\begin{array}{c} \\ R_2 \to -R_2 \\ R_3 \to R_3 + 3R_2 \end{array} \begin{pmatrix} 1 & 1 & 1 & 2 \\ 0 & 1 & 5 & 11 \\ 0 & 0 & 2t & 20 \end{pmatrix}$$

From this you see $2tz = 20 \Rightarrow z = \dfrac{10}{t}, t \neq 0$

So the required condition for solutions is that t cannot be zero.

You have from row 2: $y + 5z = 11 \Rightarrow y = 11 - \dfrac{50}{t}, t \neq 0$

... and from row 1: $x + y + z = 2 \Rightarrow x = -9 + \dfrac{40}{t}, t \neq 0$

Finally, if x is to be positive then $-9 + \dfrac{40}{t} > 0 \Rightarrow t < \dfrac{40}{9}; \quad \dfrac{40}{9} = \dfrac{440}{99}$

If y is to be positive, $11 - \dfrac{50}{t} > 0 \Rightarrow t > \dfrac{50}{11}; \quad \dfrac{50}{11} = \dfrac{450}{99}$

If x is to be positive AND if y is to be positive then $t < \dfrac{440}{99}$ AND $t > \dfrac{450}{99}$

which is impossible. QED.

Marks
    1 for method (for structured method);

    1 marks for manipulation to upper triangular form;

    1 for value of z and the condition for it to exist;

    1 for values of x and y.

    1 for proof that no all-positive set exists.

NOTES

Gaussian elimination is requested and so must be used.

A row-swap is done to take the most complicated row to the end, otherwise the parameter will be propagated throughout the solution.

A row swap would also be essential if a_{11} in a particular question turned out to be zero; or indeed a_{22} at the next step.

Make sure you answer all the parts: the solution, the condition, the proof.

3. (a)

$$\int_0^{\frac{3}{2}} \frac{x\cos^{-1}\frac{x}{3}}{\sqrt{9-x^2}}\,dx$$

$$x = 3\cos u \Rightarrow u = \cos^{-1}\frac{x}{3}$$

$$x = 3\cos u \Rightarrow dx = -3\sin u\,du$$

$$x = 0 \Rightarrow u = \cos^{-1}0 = \frac{\pi}{2}$$

$$x = \frac{3}{2} \Rightarrow u = \cos^{-1}\frac{1}{2} = \frac{\pi}{3}$$

$$\sqrt{9-9\cos^2 u} = 3\sqrt{1-\cos^2 u} = 3\sin u$$

So performing the substitution makes the integral look like:

$$\int_{\frac{\pi}{2}}^{\frac{\pi}{3}} \frac{3\cos u.u.-3\sin u\,du}{3\sin u}$$

This tidies up to

$$-3\int_{\frac{\pi}{2}}^{\frac{\pi}{3}} u\cos u\,du$$

Integrating by parts:

$$-3\left[u\sin u - \int \sin u\,du\right]_{\frac{\pi}{2}}^{\frac{\pi}{3}} = -3\left[u\sin u + \cos u\right]_{\frac{\pi}{2}}^{\frac{\pi}{3}}$$

$$= -3\left[\left(\frac{\pi}{3}\sin\frac{\pi}{3} + \cos\frac{\pi}{3}\right) - \left(\frac{\pi}{2}\sin\frac{\pi}{2} + \cos\frac{\pi}{2}\right)\right]_{\frac{\pi}{2}}^{\frac{\pi}{3}}$$

$$= -3\left(\frac{\pi}{2\sqrt{3}} + \frac{1}{2} - \frac{\pi}{2}\right)$$

$$= 0.492 \text{ (3 d.p.)}$$

Marks   **1** for dealing with the x expressions

1 for dealing with the differentials

1 for dealing with the limits

1 for integrating by parts

1 for evaluation (exact or approximate form).

NOTES   Remember that all manifestations of x have to be substituted.

including the differential, dx and the limits which are x-values.

3. (continued)

3. (b) (i) $\frac{1}{2}\big[\sin(A+B) + \sin(A-B)\big]$

$= \frac{1}{2}(\sin A \cos B + \cos A \sin B + \sin A \cos B - \cos A \sin B)$

$= \frac{1}{2}(2\sin A \cos B)$

$= \sin A \cos B$

(ii) Using the above result:

$\sin 3x \cos 5x = \frac{1}{2}\big(\sin(3x+5x) + \sin(3x-5x)\big)$

$= \frac{1}{2}\big(\sin(8x) + \sin(-2x)\big) = \frac{1}{2}\big(\sin(8x) - \sin(2x)\big)$

$\int \sin 3x \cos 5x \; dx = \frac{1}{2}\int \sin 8x - \sin 2x \; dx$

$= \frac{1}{2}\left(\frac{-\cos 8x}{8} + \frac{\cos 2x}{2}\right) + c$

$= \frac{1}{4}\cos 2x - \frac{1}{16}\cos 8x + c$

Marks **1** for proving the trigonometric identity.

1 for using the result to transform the expression to integrate.

1 for integrating.

NOTES

Formulae given in the Higher exam are expected to be known.

The first part of this question leads to a formula for converting the product of a sine and a cosine into a sum. The terms of the sum are easy to integrate.

The keyword 'Hence' tells you to use this result in the second part.

4. $z^5 = 32$

$\Rightarrow z^5 = 32\big(\cos(2\pi n) + i\sin(2\pi n)\big), n \in Z$

$\Rightarrow z = \sqrt[5]{32}\big(\cos(2\pi n) + i\sin(2\pi n)\big)^{\frac{1}{5}}$

$\Rightarrow z = 2\left(\cos\left(\frac{2\pi n}{5}\right) + i\sin\left(\frac{2\pi n}{5}\right)\right)$

using DeMoivre's theorem.

We find the set of solutions by letting $n = 0$, $n = \pm 1$, and $n = \pm 2$ giving

$z = 2(\cos 0 + i\sin 0) = 2$

$z = 2\left(\cos\frac{2\pi}{5}\right) \pm i2\left(\sin\frac{2\pi}{5}\right) = 0.618 \pm 1.902i$

$z = 2\left(\cos\frac{4\pi}{5}\right) \pm i2\left(\sin\frac{4\pi}{5}\right) = -1.618 \pm 1.176i$

Coefficients given to 3 d.p.

Marks **1** for method (expressing 1 in general form);

1 for processing … using DeMoivre's theorem correctly.

1 for simplification to the required form.

4. (continued)

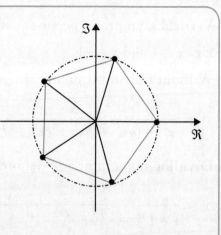

NOTES

The five solutions lies on a circle with radius 2 units and centre the origin.
They form the vertices of a regular pentagon.
You could be asked to draw this in the exam.

To get into this problem you must express 1 in the form $\cos\theta + i\sin\theta$ in its most general form viz. $\cos 2\pi n + i\sin 2\pi n$, $n \in Z$. Check that for all integer values of n this expression is 1.

One can use any integer values of n, but since the argument is defined to lie between π and $-\pi$ we find it simpler to use 0, ±1, ±2.

When you use DeMoivre's theorem in the exam, state that you are doing so.

5. (a)

$$y = \frac{x^5}{(x+1)^3(x-1)^2}$$

$$\Rightarrow \ln y = 5\ln x - 3\ln(x+1) - 2\ln(x-1)$$

$$\Rightarrow \frac{1}{y}\frac{dy}{dx} = \frac{5}{x} - \frac{3}{x+1} - \frac{2}{x-1} = \frac{x-5}{x(x+1)(x-1)}$$

$$\Rightarrow \frac{dy}{dx} = \frac{x-5}{x(x+1)(x-1)} \cdot y = \frac{x^4(x-5)}{(x+1)^4(x-1)^3}$$

Stationary points occur when $\dfrac{dy}{dx} = 0$ i.e. when $x = 0$ or $x = 5$

Create a table of signs to determine the nature of the stationary points at these values of x:

x	$\rightarrow$	0	$\rightarrow$		$\rightarrow$	5	$\rightarrow$
x^4	+	0	+		+	+	+
$(x-5)$	–	–	–		–	0	+
$(x+1)^4$	+	+	+		+	+	+
$(x-1)^3$	–	–	–		+	+	+
dy/dx	+	0	+		–	0	+
shape	/	—	/		\	–	/
nature		pt of inflexion				minimum T.P.	

Evaluating the corresponding values of y you get $(0, 0)$ is a point of inflexion on an increasing section of the curve; $(5, 0{\cdot}9)$ is a minimum turning point ... to 1 d.p.

5. (continued)

5. (*b*) Vertical asymptotes occur at values that make the denominator zero:

viz. $x = 1$ and $x = -1$.

Without fully multiplying out you see that

$$y = \frac{x^5}{x^5 + (ax^4 + bx^3 + cx^2 + dx + e)}$$

Dividing numerator and denominator by x^5 you get

$$y = \frac{1}{1 + \left(\dfrac{a}{x^1} + \dfrac{b}{x^2} + \dfrac{c}{x^3} + \dfrac{d}{x^4} + \dfrac{e}{x^5}\right)}.$$

As x tends to infinity, each fraction tends to zero and y tends to 1.
Thus $y = 1$ is a horizontal asymptote.

5. (*c*) When $x = 0.7$, $y = 0.4$; when $x = 0.8$, $y = 1.4$. Thus in the region $0.7 < x < 0.8$ the value of y attains the value 1 … and crosses the asymptote.

5. (*d*)

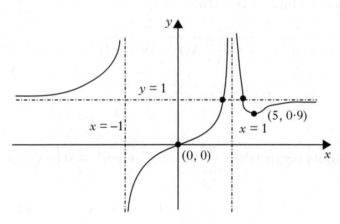

Marks

1 for preparing the function by taking the log of both sides.

1 for using the laws of logs to prepare for differentiating

1 for differentiating

1 for finding the stationary points and using a table of signs to establish the nature of these stationary points.

1 for vertical asymptotes

1 for horizontal asymptote

1 for argument leading to conclusion that curve cuts its own asymptote.

1 for a consistent sketch

NOTES

The marking scheme should be considered together for this question. It is the response taken as a whole that makes sense.

This type of function is the sort that is suitable for logarithmic differentiation. You will not always be given the hint to use the technique.

When looking for **vertical asymptotes** examine the conditions which make the denominator zero.

When examining **non-vertical asymptotes**, make sure the numerator of the rational part is of the same or lesser order than the denominator. Divide numerator and denominator by the highest power of x; consider now what happens as x tends to infinity.

6. (a) All you need do is look for a way of giving the expression a factor.

Let $n = 41$ then the expression becomes $41^2 + 41 + 41 = 41(41 + 2)$ which has a factor of 41. Thus at this value the expression is not prime. QED

Marks   **1** for citing and demonstrating a counter-example. The most obvious is 41, but it's not the only one.

> **NOTES**   A universal statement can always be proved false by citing a counter-example.

6. (b) *Read the note below for a description of the approach to this question.*

Let $n = 6$, $a = 3$ and $b = 0$ LHS $= 2.3 + 7.0 = 6 =$ RHS.

So statement true when $n = 6$.

Let $n = 7$, $a = 0$ and $b = 1$ LHS $= 2.0 + 7.1 = 7 =$ RHS.

So statement true when $n = 7$.

Suppose the statement is true for $n = k$ **and** $n = k + 1$.

Then there exists whole numbers a_1 and b_1, a_2 and b_2 such that

$k = 2a_1 + 7b_1$ and $k + 1 = 2a_2 + 7b_2$

Consider the next pair of consecutive numbers $n = k + 2$ **and** $n = k + 3$:

$k + 2 = (k) + 2 = 2a_1 + 7b_1 + 2 = 2(a_1 + 1) + 7b_1$

$k + 3 = (k + 1) + 2 = 2a_2 + 7b_2 + 2 = 2(a_2 + 1) + 7b_2$

Since both $a_1 + 1$ and $a_2 + 1$ are whole numbers then the statement is true for $n = k + 2$ and $n = k + 3$.

Thus if the statement is true for $n = k$ and $n = k + 1$ then it is true for $n = k + 2$ and $n = k + 3$. Since we know it is true for $n = 6$ and $n = 7$ then by induction it is true for all whole numbers greater than 5.

Marks   **1** for showing proposal true for $n = 6$ and $n = 7$;

1 for assuming true for $n = k$ **and** $n = k + 1$ and starting to consider $n = k + 2$ **and** $n = k + 3$;

1 for establishing that true for $n = k$ **and** $n = k + 1$ implies true for $n = k + 2$ **and** $n = k + 3$;

1 for pulling it all together in a final conclusion.

6. (continued)

> Proof by induction normally follows a particular line of argument.
>
> - Show that the proposal is true for the lowest case
> - Show that if it is true for $n = k$ then it is true for $n = k + 1$
> - Draw a conclusion pulling together all parts of the argument.
>
> However, this variety of the question requires a slight twist to this line of thinking.
>
> If you try a few examples you will see that here are really two sequences running along side-by-side.
>
> | | |
> |---|---|
> | $6 = 2.3 + 7.0$ | $7 = 2.0 + 7.1$ |
> | $8 = 2.4 + 7.0$ | $9 = 2.1 + 7.1$ |
> | $10 = 2.5 + 7.0$ | $11 = 2.2 + 7.1$ |
> | $12 = 2.6 + 7.0$ | $13 = 2.3 + 7.1$ |
>
> Instead of showing that 'if the statement is true for $n = k$ then it is true for $n = k + 1$'
>
> we show 'if the statement is true for $n = k$ **and** $n = k + 1$ then it is true for $n = k + 2$ **and** $n = k + 3$'.
>
> We thus start with showing it is true for the smallest two numbers for which we wish to prove the statement true viz. 6 and 7.
>
> Remember to prove an *existential* statement we need only find one example that works.
>
> The inspiration for this question comes from the statement in the conditions and arrangements which states, "The concept of mathematical induction may be introduced in a familiar context such as demonstrating that 2p coins and 5p coins are sufficient to pay any sum of money greater than 3p."

NOTES

7. (a) The curve cuts the x-axis where $y = 0$:

$$-2\cos\theta - 2\cos^2\theta = 0 \Rightarrow \cos\theta = 0 \text{ or } -1$$

$$\Rightarrow \theta = \frac{\pi}{2}, \frac{3\pi}{2} \text{ or } \pi$$

$$\Rightarrow x = 2, -2 \text{ or } 0$$

The curve cuts the x-axis at $(2, 0)$, $(-2, 0)$ and $(0, 0)$

The curve cuts the y-axis where $x = 0$:

$$2\sin\theta + 2\sin\theta\cos\theta = 0 \Rightarrow \sin\theta = 0 \text{ or } \cos\theta = -1$$

$$\Rightarrow \theta = 0, \pi \text{ or } \pi$$

$$\Rightarrow y = -4, \text{ or } 0$$

The curve cuts the y-axis at $(0, -4)$ and $(0, 0)$

7. (b) $\dfrac{dy}{dx} = \dfrac{dy}{d\theta} \cdot \dfrac{d\theta}{dx} = \dfrac{2\sin\theta + 4\cos\theta\sin\theta}{2\cos\theta + 2\cos 2\theta} = \dfrac{\sin\theta + \sin 2\theta}{\cos\theta + \cos 2\theta}$

(i) The gradient is zero when the numerator is zero AND the denominator is not.

$$\sin\theta + 2\sin\theta\cos\theta = 0 \Rightarrow \sin\theta = 0 \text{ or } \cos\theta = -\frac{1}{2}$$

$$\Rightarrow \theta = 0, \pi, \frac{2\pi}{3} \text{ or } \frac{4\pi}{3}$$

7.   (continued)

However, checking the value of the denominator at these four values, we find that when $\theta = \pi$ the denominator is zero.

$$\frac{dy}{dx} = 0 \Rightarrow \theta = 0, \frac{2\pi}{3} \text{ or } \frac{4\pi}{3}$$

which corresponds to the points $(0, -4)$, $\left(\dfrac{\sqrt{3}}{2}, \dfrac{1}{2}\right)$ and $\left(-\dfrac{\sqrt{3}}{2}, \dfrac{1}{2}\right)$

(ii)   The derivative is undefined when the denominator is zero:

$$\cos\theta + \cos 2\theta = 0$$
$$\Rightarrow 2\cos^2\theta + \cos\theta - 1 = 0$$
$$\Rightarrow (2\cos\theta - 1)(\cos\theta + 1) = 0$$
$$\Rightarrow \cos\theta = \frac{1}{2}, -1$$
$$\Rightarrow \theta = \frac{\pi}{3}, \frac{5\pi}{3} \text{ or } \pi$$

which corresponds to the points $\left(\dfrac{3\sqrt{3}}{2}, -\dfrac{3}{2}\right), \left(-\dfrac{3\sqrt{3}}{2}, -\dfrac{3}{2}\right)$ or $(0, 0)$

7.   (c)   The hint tells you there is a minimum at $(0, -4)$ and maxima at

$\left(\dfrac{\sqrt{3}}{2}, \dfrac{1}{2}\right)$ and $\left(-\dfrac{\sqrt{3}}{2}, \dfrac{1}{2}\right)$

Pulling all the features together you get a cardioid (heart-shaped curve)

Love and Kisses

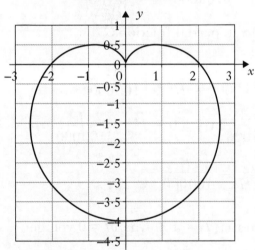

Marks   **2** for using conditions for crossing the axes;

and finding the x intercepts.

and for finding the y intercepts.

1 for method of finding derivative given parametric equations

1 for differentiating

1 for solving relevant equation and finding θ values

1 for conditions which result in undefined gradient

1 for finding the (x, y) - values.

1 for sketch which corresponds to findings.

7. (continued)

NOTES

The question is quite long and had to be curtailed for exam purposes.

Information was given to avoid you having to explore the concavity by finding the second derivative.

However, it is possible that you may have been asked:

Find $\dfrac{d^2y}{dx^2}$ and the concavity at the stationary points.

For this you would have proceeded:

$$\frac{d^2y}{dx^2} = \frac{d\left(\frac{dy}{dx}\right)}{dx} = \frac{d\left(\frac{dy}{dx}\right)}{d\theta} \cdot \frac{d\theta}{dx}$$

$$= \frac{(\cos\theta + \cos 2\theta)(\cos\theta + 2\cos 2\theta) + (\sin\theta + \sin 2\theta)(\sin\theta + 2\sin 2\theta)}{2(\cos\theta + \cos 2\theta)^3}$$

In this clumsy form you can put in the values of θ corresponding to stationary points and find their nature. The fraction does simplify to give

$$\frac{d^2y}{dx^2} = \frac{3}{2(\cos\theta + \cos 2\theta)^2} \cdot \frac{1}{(2\cos\theta - 1)}$$

$\dfrac{d^2y}{dx^2} < 0 \Rightarrow 2\cos\theta - 1 < 0 \Rightarrow \dfrac{\pi}{3} < \theta < \dfrac{5\pi}{3}$ (use a sketch of the cosine curve in the range to verify the conclusion) which corresponds to the y-values being less than $-1\cdot5$.

8. (a)

$$\frac{dH}{dt} = \frac{H}{100\,000}(7000 - H)$$

$$\Rightarrow \int \frac{dH}{H(7000 - H)} = \int \frac{dt}{100\,000}$$

Expressing the LHS in partial fractions:

$$\frac{1}{7000}\int \frac{dH}{H} + \frac{1}{7000}\int \frac{dH}{7000 - H} = \int \frac{dt}{100\,000}$$

$$\Rightarrow \frac{1}{7000}\ln|H| - \frac{1}{7000}\ln|7000 - H| = \frac{t}{100\,000} + c$$

$$\Rightarrow \frac{1}{7000}\ln\left|\frac{H}{7000 - H}\right| = \frac{t}{100\,000} + c$$

Using the condition that $H = 100$ when $t = 0$ you get

$$c = \frac{1}{7000}\ln\left(\frac{100}{6900}\right)$$

8. (continued)

$$\Rightarrow \frac{1}{7000}\ln\left(\frac{H}{7000-H}\right) = \frac{t}{100\,000} + \frac{1}{7000}\ln\left(\frac{1}{69}\right)$$

$$\Rightarrow \ln\left(\frac{H}{7000-H}\right) = \frac{7t}{100} + \ln\left(\frac{1}{69}\right)$$

$$\Rightarrow \frac{H}{7000-H} = e^{\frac{7t}{100}+\ln\left(\frac{1}{69}\right)} = e^{\frac{7t}{100}}.e^{\ln\left(\frac{1}{69}\right)} = \frac{1}{69}e^{\frac{7t}{100}}$$

$$\Rightarrow H = \frac{7000}{69}e^{0\cdot07t} - \frac{H}{69}e^{0\cdot07t}$$

$$\Rightarrow H\left(1 + \frac{e^{0\cdot07t}}{69}\right) = \frac{7000}{69}e^{0\cdot07t}$$

$$\Rightarrow H = \frac{\dfrac{7000}{69}e^{0\cdot07t}}{1 + \dfrac{e^{0\cdot07t}}{69}} = \frac{7000e^{0\cdot07t}}{69 + e^{0\cdot07t}}$$

8. (b) $H_{24} = \dfrac{7000e^{0\cdot07\times24}}{69 + e^{0\cdot07\times24}} = 505$ hectares to the nearest hectare.

8. (c) To find a value of t you are better going back to the line

$$\ln\left(\frac{H}{7000-H}\right) = \frac{7t}{100} + \ln\left(\frac{1}{69}\right) \text{ which gives you}$$

$$\ln\left(\frac{H}{7000-H}\right) - \ln\left(\frac{1}{69}\right) = \frac{7}{100}t$$

$$\Rightarrow t = \frac{100}{7}\ln\left(\frac{69H}{7000-H}\right)$$

Now 90% of 7000 is 6300.

$$t_{6300} = \frac{100}{7}\ln\left(\frac{69\times6300}{7000-6300}\right) = 92\cdot0 \text{ hours}$$

Marks **8.** (a) **1** for separating the variables

1 for using partial fractions

1 for integration

1 for simplification

1 for using the conditions and f finding c.

1 for changing the subject to H.

8. (b) **1** for finding H_{24}

8. (c) **2** for finding $t_{90\%}$

8. (continued)

9. (a) You set up AI and by a series of EROs you convert A to I. Simultaneously the same EROs will convert I to A^{-1}.

$$\left(\begin{array}{ccc} 4 & 1 & 1 \\ 1 & 0 & 2 \\ 3 & 1 & 0 \end{array}\right)\left(\begin{array}{ccc} 1 & 0 & 0 \\ 0 & 1 & 0 \\ 0 & 0 & 1 \end{array}\right)$$

$\begin{array}{l} R_1 \rightarrow R_1 / a_{11} \\ R_2 \rightarrow R_2 - a_{21}R_1 \\ R_3 \rightarrow R_3 - a_{31}R_1 \end{array}$ $\left(\begin{array}{ccc} 1 & \frac{1}{4} & \frac{1}{4} \\ 0 & -\frac{1}{4} & \frac{7}{4} \\ 0 & \frac{1}{4} & -\frac{3}{4} \end{array}\right)\left(\begin{array}{ccc} \frac{1}{4} & 0 & 0 \\ -\frac{1}{4} & 1 & 0 \\ -\frac{3}{4} & 0 & 1 \end{array}\right)$

$\begin{array}{l} R_2 \rightarrow R_2 / a_{22} \\ R_3 \rightarrow R_3 - a_{32}R_2 \end{array}$ $\left(\begin{array}{ccc} 1 & \frac{1}{4} & \frac{1}{4} \\ 0 & 1 & -7 \\ 0 & 0 & 1 \end{array}\right)\left(\begin{array}{ccc} \frac{1}{4} & 0 & 0 \\ 1 & -4 & 0 \\ -1 & 1 & 1 \end{array}\right)$

$\begin{array}{l} R_1 \rightarrow R_1 - a_{13}R_3 \\ R_2 \rightarrow R_2 - a_{23}R_3 \end{array}$ $\left(\begin{array}{ccc} 1 & \frac{1}{4} & 0 \\ 0 & 1 & 0 \\ 0 & 0 & 1 \end{array}\right)\left(\begin{array}{ccc} \frac{1}{2} & -\frac{1}{4} & -\frac{1}{4} \\ -6 & 3 & 7 \\ -1 & 1 & 1 \end{array}\right)$

$R_1 \rightarrow R_1 - a_{12}R_2$ $\left(\begin{array}{ccc} 1 & 0 & 0 \\ 0 & 1 & 0 \\ 0 & 0 & 1 \end{array}\right)\left(\begin{array}{ccc} 2 & -1 & -2 \\ -6 & 3 & 7 \\ -1 & 1 & 1 \end{array}\right)$

So $A^{-1} = \left(\begin{array}{ccc} 2 & -1 & -2 \\ -6 & 3 & 7 \\ -1 & 1 & 1 \end{array}\right)$

Marks **1** for method

1 for achieving upper triangular form

1 for achieving the identity matrix and stating inverse

9. (continued)

9. *(b)* $AB = C \Rightarrow A^{-1}AB = A^{-1}C \Rightarrow IB = A^{-1}C \Rightarrow B = A^{-1}C$

$$B = \begin{pmatrix} 2 & -1 & -2 \\ -6 & 3 & 7 \\ -1 & 1 & 1 \end{pmatrix}\begin{pmatrix} 4 & 3 & 8 \\ 9 & 5 & 1 \\ 2 & 7 & 6 \end{pmatrix}$$

$$B = \begin{pmatrix} 2.4 + (-1).9 + (-2).2 & 2.3 + (-1).5 + (-2).7 & 2.8 + (-1).1 + (-2).6 \\ -6.4 + 3.9 + 7.2 & -6.3 + 3.5 + 7.7 & -6.8 + 3.1 + 7.6 \\ -1.4 + 1.9 + 1.2 & -1.3 + 1.5 + 1.7 & -1.8 + 1.1 + 1.6 \end{pmatrix}$$

$$= \begin{pmatrix} -5 & -13 & 3 \\ 17 & 46 & -3 \\ 7 & 9 & -1 \end{pmatrix}$$

Marks   **1** for the justifying algebra

1 for matrix multiplication

NOTES > It is important to show the matrix algebra. This will help you determine whether to pre or post multiply. You do not need to show all the individual calculations. They are only shown above to be instructive.

9. *(c)* A square matrix is non-singular if its inverse exists.

The inverse of a matrix exists if its determinant is non-zero.

$$|B| = \begin{vmatrix} -5 & -13 & 3 \\ 17 & 46 & -3 \\ 7 & 9 & -1 \end{vmatrix} = -5\begin{vmatrix} 46 & -3 \\ 9 & -1 \end{vmatrix} - (-13)\begin{vmatrix} 17 & -3 \\ 7 & -1 \end{vmatrix} + 3\begin{vmatrix} 17 & 46 \\ 7 & 9 \end{vmatrix}$$

$$= -5.(-46 + 27) + 13.(-17 + 21) + 3.(153 - 322)$$

$$= -360.$$

Since this is non-zero, the matrix B has an inverse and is non-singular.

Marks   **1** for describing the conditions needed for the matrix to be non-singular.

1 for proving by demonstration that the determinant of B is non zero.

NOTES > This part is checking that you know the term non-singular and that you know how to evaluate the determinant of a 2 × 2 and a 3 × 3 matrix. Show your steps.

10. *(a)* Introduce the parameters into the equations:

$$\frac{x + 1}{2} = \frac{y + 1}{3} = \frac{z}{3} = t_2 \Rightarrow x = 2t_1 - 1; \; y = 3t_1 - 1; \; z = 3t_1$$

$$\frac{x + 8}{3} = \frac{y + 10}{4} = \frac{z + 3}{2} = t_1 \Rightarrow x = 3t_2 - 8; \; y = 4t_2 - 10; \; z = 2t_2 - 3$$

At the point of intersection you can equate corresponding expressions.

Equating x expressions: $2t_1 - 1 = 3t_2 - 8 \Rightarrow 2t_1 - 3t_2 = -7$ ...   ①

Equating y expressions: $3t_1 - 1 = 4t_2 - 10 \Rightarrow 3t_1 - 4t_2 = -9$ ...   ②

10. (continued)

Equating z expressions: $3t_1 = 2t_2 - 3 \Rightarrow 3t_1 - 2t_2 = -3$ ... ③

① × 3: $\qquad\qquad 6t_1 - 9t_2 = -21$ ... ④

② × 2: $\qquad\qquad 6t_1 - 8t_2 = -18$ ... ⑤

⑤ − ④: $\qquad\qquad\quad t_2 = 3$ ⑥

Substitute in ①: $2t_1 - 9 = -7 \Rightarrow t_1 = 1$

Check the system of equations is consistent by testing that these values satisfy equation ③ ... $3.1 - 2.3 = 3 - 6 = -3$ Equation is satisfied.

Now use $t_1 = 1$ to calculate the coordinates.

The lines intersect at $(1, 2, 3)$.

Marks

1 for re-introducing the parameters t_1 and t_2 and setting up a system of equations.

1 for solving a pair of these equations simultaneously.

1 for checking that solution satisfies the third equation.

1 for finding the point of intersection.

NOTES

This question could have been worded "Prove the lines intersect and find the point of intersection" or "Do the lines intersect?"

You can always find a pair of parameters which will satisfy two of the equations simultaneously. It is only when the parameters satisfy all three equations that intersection occurs.

Always check the third equation is satisfied by the parameters that satisfy the first two.

10. (b) The vector $\begin{pmatrix} 2 \\ 3 \\ 3 \end{pmatrix}$ is parallel to the first line and $\begin{pmatrix} 3 \\ 4 \\ 2 \end{pmatrix}$ is parallel to the second.

To find a vector normal to both lines we find their cross product.

$$\begin{pmatrix} 2 \\ 3 \\ 3 \end{pmatrix} \times \begin{pmatrix} 3 \\ 4 \\ 2 \end{pmatrix} = \begin{vmatrix} \boldsymbol{i} & \boldsymbol{j} & \boldsymbol{k} \\ 2 & 3 & 3 \\ 3 & 4 & 2 \end{vmatrix} = \boldsymbol{i}\begin{vmatrix} 3 & 3 \\ 4 & 2 \end{vmatrix} - \boldsymbol{j}\begin{vmatrix} 2 & 3 \\ 3 & 2 \end{vmatrix} + \boldsymbol{k}\begin{vmatrix} 2 & 3 \\ 3 & 4 \end{vmatrix}$$

$$= \boldsymbol{i}(6 - 12) - \boldsymbol{j}(4 - 9) + \boldsymbol{k}(8 - 9)$$

$$= -6\boldsymbol{i} + 5\boldsymbol{j} - \boldsymbol{k}$$

$$= \begin{pmatrix} -6 \\ 5 \\ -1 \end{pmatrix}$$

This is also a normal to the required plane.

So the plane will have an equation of the form $\begin{pmatrix} -6 \\ 5 \\ -1 \end{pmatrix}\begin{pmatrix} x \\ y \\ z \end{pmatrix} = k$ where k is a constant.

10. (continued)

We know that (1, 2, 3) lies on both lines and hence on the plane.

So $\begin{pmatrix} -6 \\ 5 \\ -1 \end{pmatrix}\begin{pmatrix} 1 \\ 2 \\ 3 \end{pmatrix} = k \Rightarrow k = -6 + 10 - 3 = 1$

So the equation of the plane is $\begin{pmatrix} -6 \\ 5 \\ -1 \end{pmatrix}\begin{pmatrix} x \\ y \\ z \end{pmatrix} = 1$

i.e. $-6x + 5y - z = 1$

Marks

1 for finding a suitable normal

1 for finding the constant k.

1 for communicating required equation

> *NOTES*
>
> To get the equation of a plane we require
>
> (i) a normal to the plane.
>
> (Usually the cross product of two vectors on the plane will do.)
>
> (ii) a point on the plane.
>
> (In this case, a point on both lines that lie on the plane.)

10. (c) The vector $\boldsymbol{u} = \begin{pmatrix} -6 \\ 5 \\ -1 \end{pmatrix}$ is parallel to the required line. A(0, 0, 0) lies on the line.

Given that the typical point P, on the line is (x, y, z), the equation of the line is $\overrightarrow{AP} = t\boldsymbol{u}$ where t is a parameter.

The required equation is $\begin{pmatrix} x - 0 \\ y - 0 \\ z - 0 \end{pmatrix} = \begin{pmatrix} -6t \\ 5t \\ -t \end{pmatrix}$

Expressed as a system of parametric equations we get:

$x = -6t; \ y = 5t; \ z = -t$.

This will cut the plane $-6x + 5y - z = 1$ when $-6(-6t) + 5(5t) - (-t) = 1$

$\Rightarrow 62t = 1 \Rightarrow t = \dfrac{1}{62}$

If you had been asked for the point of intersection …

$(-6t, 5t, -t) = \left(-\dfrac{6}{62}, \dfrac{5}{62}, -\dfrac{1}{62} \right)$

Marks

1 for pulling together the facts required to form the equation viz. the parallel vector and the point [implied].

1 for expressing equation in the required form.

and substitution leading to the value of t.

> *NOTES*
>
> The working here is expanded because it is meant to be explanatory. One would not need to go through it all; the three lines starting with "The required equation is …" would be enough.

10. (continued)

10. (d) Using the definition of the dot product in the form $\cos\theta = \dfrac{\boldsymbol{u}.\boldsymbol{v}}{|\boldsymbol{u}||\boldsymbol{v}|}$ and the unit vectors:

$$\cos\theta_x = \frac{\begin{pmatrix} -6 \\ 5 \\ -1 \end{pmatrix}\begin{pmatrix} 1 \\ 0 \\ 0 \end{pmatrix}}{\sqrt{62}\sqrt{1}} = \frac{-6}{\sqrt{62}} \quad \text{where } \theta_x \text{ is the angle the line makes with the}$$

x-axis.

$$\text{Similarly } \cos\theta_y = \frac{\begin{pmatrix} -6 \\ 5 \\ -1 \end{pmatrix}\begin{pmatrix} 0 \\ 1 \\ 0 \end{pmatrix}}{\sqrt{62}\sqrt{1}} = \frac{5}{\sqrt{62}} \quad \text{and} \quad \cos\theta_z = \frac{\begin{pmatrix} -6 \\ 5 \\ -1 \end{pmatrix}\begin{pmatrix} 0 \\ 0 \\ 1 \end{pmatrix}}{\sqrt{62}\sqrt{1}} = \frac{-1}{\sqrt{62}}$$

Marks **1** for method … using basis vectors.

NOTES These cosines are often called *direction cosines* as they can be used to describe the direction the vector takes. The angles the line makes with the axes can readily be obtained … If you work in degrees we get $\theta_x = 139.6$, $\theta_y = 50.6$, $\theta_z = 97.3$.

11. (a) $\displaystyle\sum_{r=1}^{n} \frac{1}{r^2 + r}$

Marks **1** for using sigma notation correctly.

11. (b) $\dfrac{1}{r^2 + r} = \dfrac{1}{r(r+1)} = \dfrac{1}{r} - \dfrac{1}{r+1}$

$$\Rightarrow \sum_{r=1}^{n} \frac{1}{r^2 + r} = \sum_{r=1}^{n} \frac{1}{r(r+1)} = \sum_{r=1}^{n} \frac{1}{r} - \sum_{r=1}^{n} \frac{1}{r+1}$$

$$\frac{1}{R_n} = \left(\frac{1}{1} + \frac{1}{2} + \frac{1}{3} + \cdots + \frac{1}{n} \right) - \left(\frac{1}{2} + \frac{1}{3} + \frac{1}{4} + \cdots + \frac{1}{n} + \frac{1}{n+1} \right)$$

$$\frac{1}{R_n} = 1 - \frac{1}{n+1} = \frac{n}{n+1}$$

$$\Rightarrow R_n = \frac{n+1}{n}$$

Marks **1** for partial fractions.

1 for interpretation and expansion.

1 for elimination of terms, simplification and finding R_n.

NOTES Whenever you get $\displaystyle\sum_{r=1}^{n} f(r) - \sum_{r=1}^{n} f(r + a)$ you can simplify it by expansion and elimination.

Remember to answer the exact question that was asked.

11. (continued)

11. (c)
$$\sum_{r=2}^{n} \frac{1}{r^2 - 1} = \sum_{r=2}^{n} \frac{1}{(r-1)(r+1)} = \frac{1}{2}\sum_{r=2}^{n} \frac{1}{r-1} - \frac{1}{2}\sum_{r=2}^{n} \frac{1}{r+1}$$

$$= \frac{1}{2}\left(\frac{1}{1} + \frac{1}{2} + \frac{1}{3} + \cdots + \frac{1}{n-1}\right) - \frac{1}{2}\left(\frac{1}{3} + \frac{1}{4} + \frac{1}{5} + \cdots + \frac{1}{n-1} + \frac{1}{n} + \frac{1}{n+1}\right)$$

$$= \frac{1}{2}\left(1 + \frac{1}{2} - \frac{1}{n} - \frac{1}{n+1}\right) = \frac{3n^2 - n - 2}{4n^2 + 4n}$$

$$= \frac{3}{4} - \frac{4n + 2}{4n^2 + 4n}$$

$$= \frac{3}{4} - \frac{\frac{4}{n} + \frac{2}{n^2}}{4 + \frac{4}{n}}$$

As n tends to infinity, each fraction with a denominator n or n^2 tends to zero, and thus the sum tends to $\frac{3}{4}$.

Marks

1 for partial fractions.

1 for interpretation and expansion, elimination of terms, simplification.

1 for considerations leading to a limit.

> **NOTES**
>
> When examining the behaviour of a rational function
>
> (i)   make sure that the numerator is of a lower degree than the denominator, by division if necessary;
>
> (ii) divide each term in the rational part by the highest power of n.
>
> (iii) consider what happens to each part as n tends to infinity,
>
> In this case $\dfrac{3}{4} - \dfrac{0 + 0}{4 + 0} = \dfrac{3}{4}$

12. Divide throughout to make the coefficient of the $\dfrac{dy}{dx}$ term unity.

$$\frac{dy}{dx} + y\tan x = x\sec x$$

The integrating factor, $I = e^{\int \tan x \; dx} = e^{-\ln(\cos x)} = e^{\ln(\cos x)^{-1}} = (\cos x)^{-1} = \sec x$

Multiply throughout by the integrating factor

$$\frac{dy}{dx}\sec x + y\sec x \tan x = x\sec^2 x$$

$$\Rightarrow \frac{d(y\sec x)}{dx} = x\sec^2 x$$

12. (continued)

Now you can integrate both sides:

$$y\sec x = \int x\sec^2 x \ dx$$

$$\Rightarrow y\sec x = x\tan x - \int \tan x \ dx$$

$$\Rightarrow y\sec x = x\tan x + \ln(\cos x) + c$$

$$\Rightarrow y = x\sin x + \cos x\ln(\cos x) + c\cos x$$

If $y = 1$ when $x = 0$ then

$$1 = 0.\sin 0 + \cos 0.\ln(\cos 0) + c\cos 0$$

$$\Rightarrow 1 = 0 + 1.\ln 1 + c$$

$$\Rightarrow c = 1$$

The particular solution is $y = x\sin x + \cos x\ln(\cos x) + \cos x$

Marks   **1** for manipulating the equation to a standard form.

1 for finding the integrating factor.

1 for multiplying by I and preparing to integrate.

1 for integrating.

1 for making y the subject of the equation.

1 for finding c and the particular solution.

NOTES

Remember $\int \tan x \ dx = \int \dfrac{\sin x}{\cos x}dx = -\ln(\cos x)$

Integration by parts has been used to find $\int x\sec^2 x \ dx$

13. (*a*) $x = \cos^3 \theta \Rightarrow dx = -3\cos^2 \theta \sin\theta \, d\theta$

(*b*) $\dfrac{dy}{dx} = \dfrac{dy}{d\theta} \div \dfrac{dx}{d\theta} = \dfrac{3\sin^2 \theta \cos\theta}{-3\cos^2 \theta \sin\theta} = -\tan\theta$

(*c*)   First find the length of the part in the first quadrant, i.e. from $x = 0$ to $x = 1$.

This corresponds to the range $\theta = \dfrac{\pi}{2}$ to $\theta = 0$

$$\int_a^b \sqrt{1 + \left(\frac{dy}{dx}\right)^2}\, dx = \int_{\frac{\pi}{2}}^0 \sqrt{1 + (-\tan\theta)^2} \cdot -3\cos^2 \theta \sin\theta \, d\theta$$

$$= -3\int_{\frac{\pi}{2}}^0 \sqrt{\sec^2 \theta} \cdot \cos^2 \theta \sin\theta \, d\theta = -3\int_{\frac{\pi}{2}}^0 \cos\theta \sin\theta \, d\theta = -\frac{3}{2}\int_{\frac{\pi}{2}}^0 \sin 2\theta \, d\theta$$

$$= -\frac{3}{2}\left[-\frac{1}{2}\cos 2\theta\right]_{\frac{\pi}{2}}^0 = -\frac{3}{2}\left(-\frac{1}{2} - \frac{1}{2}\right) = \frac{3}{2}$$

The whole perimeter is four times this amount viz. 6 units.

13. (continued)

Marks

(a) **1** for finding required expression

(b) **1** for finding $\dfrac{dy}{d\theta}$ and $\dfrac{dx}{d\theta}$

   and combining them to find $\dfrac{dy}{dx}$

(c) **1** for substitutions

   1 for simplification to form which can be integrated directly.

   1 for integration and substitution of limits

   1 for evaluation and perimeter.

> **NOTES**
>
> There are many applications of integration, and thus many formulae which involve integration. If you are required to apply integration you will probably be given the formulae ... the exceptions being related to speed, distance, time and to volumes of revolution.
>
> If parametric equations are involved, one must remember that if the derivative is expressed in terms of the parameter then so must the differential dx. The limits of the integration must also be considered as parameter values unless x is reintroduced prior to substitution.
>
> And finally, you are expected to remember the trigonometric formulae you learned in the Higher Course, including compound angle and double angle formulae.

14. (a) The GCD can be found using the Euclidean algorithm.

$40152 = 1.31752 + 8400$

$31752 = 3.8400 + 6552$

$8400 = 1.6552 + 1848$

$6552 = 3.1848 + 1008$

$1848 = 1.1008 + 840$

$1008 = 1.840 + 168$

$840 = 5.168 + 0$

168 is the last non-zero remainder.

168 is the GCD of C and V

The interval between photos is 168 hours.

Marks **1** for identifying the GCD as the required interval

   1 for using the Euclidean algorithm

   1 for the manipulations and the final interval.

> **NOTES** Communication is important here and a good layout is required to help with later manipulations.

14. (b) You are given $L(a, b) = \dfrac{ab}{G(a, b)}$

$$L(40152, 31752) = \frac{40\ 152 \times 31\ 752}{G(40\ 152, 31\ 752)} = \frac{1\ 274\ 906\ 304}{168} = 7\ 588\ 728$$

The same positions will be attained at common multiples of C and V.

The first time will be at the lowest common multiple viz. 7 588 728 hours.

14. (continued)

Marks **1** for finding the LCM and communicating the answer in context.

> **NOTES** Interpretation is important here and identifying the LCM should be a feature of your answer. Being seen to use the given formula is also important.
>
> Make use of the Euclidean algorithm.

14. (c) Using the working from part (a)

$$168 = 1008 - 1.840$$
$$= 1008 - 1.(1848 - 1.1008) = 2.1008 - 1.1848$$
$$= 2.(6552 - 3.1848) - 1.1848 = 2.6552 - 7.1848$$
$$= 2.6552 - 7.(8400 - 1.6552) = 9.6552 - 7.8400$$
$$= 9(31752 - 3.8400) - 7.8400 = 9.31752 - 34.8400$$
$$= 9.31752 - 34.(40512 - 1.31752) = 43.31752 - 34.40152$$
$$168 = (-34)C + 43V$$

Using $L(a, b) = \dfrac{ab}{G(a, b)}$

$$\frac{1}{L(C,V)} = \frac{G(C,V)}{CV} = \frac{-34C}{CV} + \frac{43V}{CV}$$

$$\Rightarrow \frac{1}{L(C,V)} = \frac{-34}{V} + \frac{43}{C}$$

Marks **1** Reversing the steps of the Euclidean algorithm

1 Using the identity

1 Coming to the required conclusion

> **NOTES** Again communication is critical and steps should be clearly explained.

15. The auxiliary equation: $m^2 - 6m + 13 = 0$

$$\Rightarrow m = \frac{6 \pm \sqrt{36 - 52}}{2} = 3 \pm 2i$$

So the complementary function (CF) $= e^{3x}(A\cos 2x + B\sin 2x)$

For a particular integral (PI) try $y = a\sin 2x + b\cos 2x$

$$\Rightarrow \frac{dy}{dx} = 2a\cos 2x - 2b\sin 2x$$

$$\Rightarrow \frac{d^2y}{dx^2} = -4a\sin 2x - 4b\cos 2x$$

$$\frac{d^2y}{dx^2} - 6\frac{dy}{dx} + 13y - 21\sin 2x = 0$$

$$\Rightarrow -4a\sin 2x - 4b\cos 2x - 12a\cos 2x + 12b\sin 2x + 13a\sin 2x + 13b\cos 2x = 21\sin 2x$$

Equating coefficients of $\sin 2x$: $9a + 12b = 21$ ...   ①

Equating coefficients of $\cos 2x$: $-12a + 9b = 0$ ...   ②

15. (continued)

$4/3 \times \text{①}: 12a + 16b = 28 \qquad \ldots \qquad$ ③

$\text{③} + \text{②}: 25b = 28 \Rightarrow b = 1 \cdot 12 \qquad$ ④

Substituting into ②: $-12a + 10 \cdot 08 = 0 \Rightarrow a = 0 \cdot 84$

General Solution: $y = \text{CF} + \text{PI}$

i.e. $y = e^{3x}(A\cos 2x + B\sin 2x) + 0 \cdot 84\sin 2x + 1 \cdot 12\cos 2x$

Marks

1 for solving auxiliary equation.

1 for stating CF.

1 for choosing form for suitable PI and substituting.

1 for finding PI **and** general solution.

NOTES

When the roots of the auxiliary equation are of the form $a \pm ib$ the complementary function is $e^{ax}(A\cos bx + B\sin bx)$.

When the constant term is $p\sin qx$, try $y = e\sin qx + f\cos qx$.